INFORMATION PROCESSING AND HUMAN-MACHINE INTERACTION

AN APPROACH TO COGNITIVE ENGINEERING

North-Holland Series in
SYSTEM SCIENCE AND ENGINEERING
Andrew P. Sage, *Editor*

1	**Wismer and Chattergy**	Introduction to Nonlinear Optimization: A Problem Solving Approach
2	**Sutherland**	Societal Systems: Methodology, Modeling, and Management
3	**Šiljak**	Large-Scale Dynamic Systems
4	**Porter et al.**	A Guidebook for Technology Assessment and Impact Analysis
5	**Boroush et al.**	Technology Assessment: Creative Futures
6	**Rouse**	Systems Engineering Models of Human–Machine Interaction
7	**Sengupta**	Decision Models in Stochastic Programming: Operational Methods of Decision Making Under Uncertainty
8	**Chankong and Haimes**	Multiobjective Decision Making: Theory and Methodology
9	**Jamshidi**	Large Scale Systems: Modeling and Control
10	**Sage**	Economic Systems Analysis
11	**Fraser and Hipel**	Conflict Analysis: Models and Resolutions
12	**Rasmussen**	Information Processing and Human–Machine Interaction: An Approach to Cognitive Engineering

INFORMATION PROCESSING AND HUMAN-MACHINE INTERACTION

AN APPROACH TO COGNITIVE ENGINEERING

Series Volume 12

Jens Rasmussen
Risö National Laboratory
Roskilde, Denmark

North-Holland
New York • Amsterdam • London

Elsevier Science Publishing Co., Inc.
52 Vanderbilt Avenue, New York, New York 10017

Distributors outside the United States and Canada:

Elsevier Science Publishers B.V.
P.O. Box 211, 1000 AE Amsterdam, The Netherlands

Library of Congress Cataloging in Publication Data

Rasmussen, Jens, 1926–
Information processing and human-machine interaction.
(North-Holland series in system science and engineering; 12)

Bibliography: p.
Includes index.
1. Man-machine systems. 2. Human engineering.
3. Process control. I. Title. II. Series.
TA167.R37 1986 620.8′2 86-13599
ISBN 0-444-00987-6

Current printing (last digit);
10 9 8 7 6 5 4 3 2 1

Manufactured in the United States of America

Contents

Preface

The contents of this book have evolved as the result of many years of research concerning the design of control and safety systems for industrial process plants. The focus of this research has been the analysis of risk and reliability, particularly as related to the nuclear power and chemical process industries. In the early 1960s, we realized from analyses of industrial accidents the need for an integrated approach to the design of human–machine systems. However, we very rapidly encountered great difficulties in our efforts to bridge the gap between the methodology and concepts of control engineering and those from various branches of psychology. Because of its kinship to classical experimental psychology and its behavioristic claim for the exclusive use of objective data representing overt activity, the traditional human factors field had very little to offer.

Modern industrial control systems use advanced information technology for support of human decision making during supervisory control tasks and emergency management. Models of human information-processing abilities and limitations are prerequisites for the basic conceptual design of such systems. The experience with the early process computer installations—old boys in the trade will remember the discussions around the experimental system at the "Little Gipsy" power plant—clearly indicated the pressing need for analysis of operators' cognitive tasks and guidelines for interface design. Because of the slow transition in psychological research away from behaviorism and the preoccupation of the early research in artificial intelligence with games and theorem proving, we found it impossible to wait for guidelines from such sources. It appeared to be necessary to start our own selective research program to find models useful for engineering design. From analysis of many hours of verbal protocols recorded during real work situations in workshops and control rooms and of hundreds of incident reports (in particular, from nuclear power plants), we developed a conceptual frame of reference that served us well in formulating our problems in concepts that could be related to control system design. This framework is the topic of the book. Furthermore, the conceptual framework that we developed turned out to be very useful for relating our system design problems to discussions in the scientific literature in general, not only in the now rapidly evolving field of cognitive psychology, but also in linguistics, semiotics, biology, and philosophy. Thus, throughout this book are comments and references to key books and papers in these related areas.

The aim of this work is not to propose specific design methods or solutions for human–machine systems in a particular technical field. Technological development would rapidly outdate such a proposal. Instead, the book presents an analysis of the concepts and domains of description that can be used as a structure for the formulation of problems and development of design guidelines in the individual fields of application. I hope that the book will be useful both as a guide for control engineers and human–machine system designers, as well as a demonstration for researchers in several humanistic disciplines of the needs and problems in modern system design.

Writing in an interdisciplinary field raises many questions of terminology, particularly when not using one's native language. Most of the important concepts used in the book, such as knowledge, sign, and model, will have very specific but different connotations, depending on the individual reader's professional background. Consequently, I have tried to use terms in their commonsense and neutral meanings, hoping that the intended meaning will appear from the context irrespective of the individual reader's background. The various topics and concepts are approached several times from different points of view in a way that is intended to be convenient for those readers who are interested only in reading parts of the text.

Acknowledgments

The work presented in this book has been possible only because technicians and engineers from the Electronics Department at Risö, operators from Risö's research reactors, and operators from the Danish Vestkraft and Isefjordværket fossil power plants agreed to serve as active participants in our analyses of their work situations. Their interest and patience during recording of verbal protocols and subsequent discussions are gratefully acknowledged.

The work clearly also depends on the efforts of the entire man–machine group at Risö, in particular, L.P. Goodstein, M. Lind, O.M. Pedersen, and E. Hollnagel. Their cooperation is gratefully acknowledged, as is also the language revision and word processing assistance of Ms. Bitten Skaarup. Also, the influence of discussions in several international circles is gratefully acknowledged. Mention should be made of talks with Neville Moray, William B. Rouse, and Thomas B. Sheridan during the various NATO meetings and workshops on supervisory control and human errors, as well as with John Gallagher, Lewis F. Hanes, and Richard W. Pew during several workshops on disturbance analysis systems in nuclear power plants, sponsored by the IEEE and U.S. Nuclear Regulatory Commission. In addition, discussions during various workshops of the OECD committee on the safety of nuclear installations (CSNI), in particular, with Jacques Leplat, have been very helpful. The material in the book has been organized and used as lecture notes for a course in human–machine interaction during my stay as visiting professor at the Center for Man–Machine Systems Research, Georgia Institute of Technology, Atlanta.

I want to thank the following journals and publishers for their kind permission to use previously published material and to reproduce figures:

Chapter 5 is based on material that was first published by Francis and Taylor in *Ergonomics* 17(3), 1974.

Chapter 6 is based on my contribution to *Human Detection and Diagnosis of System Failures,* J. Rasmussen and W.B. Rouse, eds., published in 1981 by Plenum Press.

Some of the material used in Chapter 8 also appeared in my contribution to *Human Interaction with Computers,* H.T. Smith and T.R.G. Green, eds., published in 1981 by Academic Press.

Chapters 8, 9, and 10 contain material published by IEEE in *Transactions on Systems, Man, and Cybernetics,* SMC-13, No. 3, 1983.

A preliminary version of Chapters 2–7 was published in 1984 in *Advances in Man–Machine System Research, Vol. 1,* W.B. Rouse, ed., by J.A.I. Press.

Finally, I thank all publishers cited for their kind permission to reproduce figures, as mentioned in the individual captions.

INFORMATION PROCESSING AND HUMAN-MACHINE INTERACTION

AN APPROACH TO COGNITIVE ENGINEERING

Chapter 1

Introduction

Two major trends in technological development have influenced the problems that are faced by designers of the human-machine interfaces of major modern systems. One is a general trend toward larger and more complex systems with centralized control. Very often, large systems imply large concentrations of data, of energy, or of hazardous material, and maloperation can therefore have serious consequences not only for the system itself and its users, but frequently also for the environment and the general public. In spite of the introduction of automatic safety systems, industrial accidents do occur, and analyses after the fact typically find significant contribution from errors by operators. Recently, a number of major incidents, such as at Three Mile Island and Bhopal, have caused general concern for the operators' role and the efforts to support them during complex disturbances. In centralized and automated systems, the problem is that quite normal and to-be-expected human flaws may cause drastic losses. For a secretary, a slip may result in a lost manuscript on a floppy disk rather than in merely a typing error. In Michigan in 1973, an otherwise reasonable improvisation for red bags that were out of stock in a chemical company supplying cattle feed to stores resulted in widespread contamination by PBB and the need for large-scale killing of poisoned cattle.

The other major trend is the rapid development of modern information technology based on inexpensive but powerful computers. Computer-based control systems and interfaces for human-machine communication in centralized management decision systems and industrial production installations will change the work situation of decision makers and system operators in several different ways. The allocation of tasks to people and to automatic systems will change. Automation does not remove humans from the system; basically it moves them from the immediate control of system operation to higher-level supervisory tasks and to longer-term maintenance and planning tasks. That is, automation affects the general work organization. In general, this also changes the requirements for operator training, partly because the task is changed to supervisory decision making in potentially risky situations, but also because the requirements for know-how during such situations may not be met by the skill developed during more normal operating periods. Furthermore, computer-based interface systems result in new ways of preprocessing measured data and of interactive decision making that can be matched to various situations and different operator tasks. These affect the design and administration of operating instructions and manuals.

In this way, extensive use of computer-based information affects several aspects of system design that are typically considered separately, by different persons and at different phases of system design. However, these aspects are so intimately related that it is difficult to envisage their being introduced gradually and independently in an optimal way. Introducing new technology, which affects several basic factors in system design, is a kind of multidimensional optimization process in a multipeaked landscape in which the different summits represent designs based on different basic concepts. Considering that the traditional technology–for instance, the one-sensor-one-indicator technology in industrial control rooms–has been subject to designers' experiments with improvements for half a century or more, it may be assumed that an optimum has been reached. This optimum is probably rather flat, owing to the adaptability of humans and the opportunities left them for adaptation in the traditional systems. An extensive use of information technology will, it is hoped, create a new and higher peak, which will very probably be more narrow because of a higher system complexity and more intricate human-machine interaction.

Unfortunately, this peak cannot be reached in an optimal way by cautious changes of the traditional design along one dimension at a time, but only by a jump in all dimensions simultaneously. Furthermore, we can expect to hit the peak within a reasonable vicinity of the top only if we know where to look for it from the outset, i.e., if we have a conceptual framework to guide the jump.

The objective of this book is a discussion in some detail of such a framework related to analysis and description of human-machine interaction and design of interface systems. Use of computer-based information technology to support decision making in supervisory systems control necessarily implies an attempt to match the information processes of the computer to the mental decision processes of an operator. This approach does not imply that computers should process information in the same way as humans would. On the contrary, the processes used by computers and humans will have to match different resource characteristics. However, to support human decision making and supervisory control, the results of computer processing must be communicated at appropriate steps of the decision sequence and in a form that is compatible with the human decision strategy. Therefore, the designer has to predict, in one way or another, which decision strategy an operator will choose. If the designer succeeds in this prediction, a very effective human-machine cooperation may result; if not, the operator may be worse off with the new support than he or she was in the traditional system; hence, the narrow peak in the optimization.

In Chapter 2, a framework for *cognitive task analysis* will be discussed. Traditionally, task analysis is discussed in terms of the external activities required by the system. This may be due partly to a behavioristic point of view in which only descriptions in terms of observable events are professionally acceptable; and partly to the limited need for description of mental activities in relation to systems of moderate levels of automation that are based mostly on presentation of basic data. What is needed is a framework for cognitive task analysis that enables the description of a decision task in terms of the necessary information processes. This framework or model must represent human performance in various situations and at different levels of training. Such a task analysis in information processing terms can also be used to characterize information processing by computers.

In Chapter 3, the framework will be used to relate the decision task to the control requirements of the system during various situations in order to identify the content of the information processes required. In particular, the *diagnostic task* in industrial process control will be considered in detail as an example.

For system design, we are not necessarily looking for predictive models of the detailed information processes, as they will emerge in a specific situation. Rather, we need predictive models of categories of information processes that enable the prediction of that category that will be activated by a particular interface configuration and its display formats. The model will then support the choice of an interface design that can activate that category of behavior that has limiting properties compatible with the functions allocated the human operator, and for which the design has been optimized.

Two aspects of such a categorical model of the information processes are discussed in detail with reference to the diagnostic task. In Chapter 4, the knowledge of the functional or causal properties of a physical system that is necessary for making control decisions is discussed in terms of an *abstraction hierarchy*, similar to the hierarchies used for organization of data bases for CAD/CAM (computer-aided design/manufacturing) systems. In Chapter 5, the different *strategies* that can be used for state identification and diagnosis in systems control are discussed, together with their requirements for system knowledge and processing capacity.

In order to predict the strategy that an operator will choose in a specific situation, it is necessary to know his or her subjective goals and *performance criteria*. This aspect is discussed in Chapter 6 with reference to analysis of the performance of technicians in a real-life task, based on verbal protocols.

In Chapter 7, the use of the conceptual *framework for systems design* is considered. The design is formalized in a sequence of decisions, and how the implementation-independent formulation of the information processes obtained in the cognitive task analysis is used for human-computer task allocation is discussed. This task allocation depends on a systematic demand-resource matching, which results in a cooperative decision making in which the designer, the system user, and the computer are sharing roles.

The task allocation based on demand-resource matching depends on models of human resources for information processing that are discussed in the remaining part of the book. In Chapter 8, the scope of the *model of a human* information processor *viewed as a system component* is discussed and reference to selected trends in psychological research is made. In Chapter 9, aspects of *human information processing* of particular importance for the decision making implied in systems control are considered in more detail. Different domains of human behavior are identified with respect to the internal control, and the semiotic interpretations of information and actions exchanged across the human-machine interface are considered.

In Chapter 10, the abstraction hierarchy that was used for cognitive task analysis and for normative task design in Chapter 4 is reconsidered as a framework for describing the actual content of *mental models* during problem solving. In addition, the form of mental models in terms of commonsense causal reasoning and formal networks of relations are considered.

Consideration of human errors is important; successful simulation of the performance of well-trained personnel is no proof of the quality of a model. Only acceptable simulation of improper performance when requirements are increased beyond capability limits will qualify a model for use in system design. In Chapter 11, *human errors* are discussed in some detail. A discussion of their definition leads to the view that human errors are the manifestation of human variability, which is a basic part of adaptation and learning. Therefore, human errors should be met by a design of "friendly systems" rather than by attempts to avoid them by "better" training and instruction. As a basis for system design, the mechanisms behind different categories of errors are discussed with reference to the categories of internal control of behavior considered in Chapter 9.

Human performance during stress is an important problem for system design because stress can drastically influence the resources for rational thinking and, hence, the likelihood of error. A phenomenon called "tunnel vision" is often used to explain human errors during accident conditions. However, a high degree of stress is normally correlated with unfamiliar tasks that call for more selective attention. Therefore, a section is included in Chapter 9 in which the influence of unfamiliarity of tasks and of arousal owing to stress is discussed and briefly related to current research on the relation between emotion and cognition.

Different kinds of models of human information processing are suited for representations of behavior depending upon the internal mode of control. In Chapter 12, a number of available models are reviewed, ranging from mathematical models of sensorimotor behavior of highly trained people in manual control tasks, to artificial-intelligence-type models of problem-solving performance.

Chapter 2

A Framework for Cognitive Task Analysis

The design of human-machine interface systems based on modern information technology should be based on a generic model of the information processes implied in the decisions to be taken by the controller. To be generally applicable, this model must be expressed in terms independent of the specific system and its immediate control requirements. Unfortunately, in spite of a long tradition of research in management decision making, a generic model of the information processes implied does not exist. At best, the existing models can be considered normative models suited for initialization of novices (Dreyfus and Dreyfus, 1980; Mintzberg, 1973). In the present context, we have therefore chosen to benefit from the clear and well-structured nature of industrial process systems, in basing our discussion on models of supervisory control decisions for such systems.

For control of a physical system such as an industrial process plant, a normative model of the necessary decision phases can be suggested. This sequence has been developed from analysis of decision making in a power plant control room and includes the following phases. First, the decision maker has to *detect* the need for intervention and has to look around and to *observe* some important data in order to have direction for subsequent activities. He or she then has to analyze the evidence available in order to *identify* the present state of affairs, and to *evaluate* their possible consequences with reference to the established operational goals and company policies. Based on the evaluation, a *target state* into which the system should be transferred is chosen, and the *task* that the decision maker has to perform is selected from a review of the resources available to reach the target state; examples of such target states and related tasks will be discussed in the following chapter. When the task has thus been identified, the proper *procedure*, i.e., how to do it, must be planned and *executed*.

It will be noticed that the nature of the information processes changes during the sequence. In the beginning, it is an analysis of the situation to identify the problem, then a prediction and value judgment and, finally, a selection and planning of the proper control actions. Depending upon the context in which the control decisions have to be made, in particular whether it is control of a physical system governed by causal physical laws, or a social system governed by human intentions, the breakdown of the decision task may result in different elementary phases. Those used here have been chosen in order to have a flexible framework for description of the decision making found in verbal protocols from actual operation of a physical system, i.e., an industrial process plant (Rasmussen, 1976).

The protocols were recorded in an attempt to analyze the operators' information processes during plant operation. The scene was the control room of a fossile fuel power plant during a start-up by a team of highly skilled operators. As it turned out, the protocols did not, in general, reflect the operators' information processes, but a sequence of "states of knowledge" representing what the operators knew about the operating conditions of the plant and about their task performance. These states of knowledge appeared as standardized nodes between the information processes shown in Figure 2.1. The information processes themselves were only reflected by the protocols when the operators had to cope with less familiar situations. This is to be expected with a highly skilled team during a familiar situation.

The existence of break points in the information processes at standardized nodes between subroutines like those of Figure 2.1 appears to be very efficient for a variety of reasons:

The information processes of the different phases have different structures in that they are related to analysis, evaluation, and planning in various functional frameworks, i.e., they are based on different information-processing models. Therefore, a standardized "state of knowledge" that separates them serves for recoding, by being the output state of the preceding process and serving as the input state for the following process.

Break points at standardized nodes make it possible to generate a sequence for special situations by chaining subroutines of general applicability and using solutions from prior experience. This is extensively done by skilled operators, leading to a great repertoire of shortcuts and bypasses in the decision process, as illustrated in Figure 2.1.

Finally, standardized nodes are likely to be developed for easy communication among several operators sharing information and experience and cooperating in execution of the control tasks.

In addition to being a framework for representing the decision making of process plant operators, the "decision ladder" of Figure 2.1 is well suited for man-machine interface design. The identification of the standardized nodes linking the elementary information processes also makes it possible to plan in a

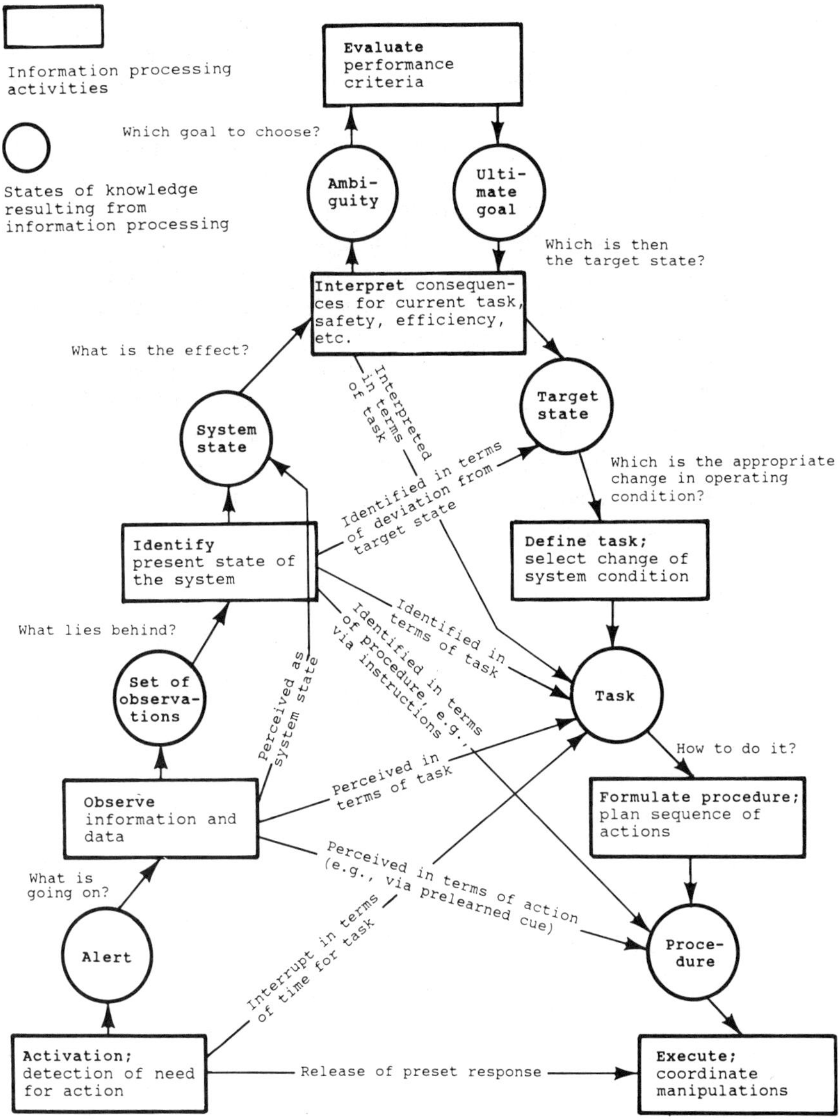

Figure 2.1. Schematic map of the sequence of information processes involved in a control decision. Rational, causal reasoning connects the "states of knowledge" in the basic sequence. Stereotyped processes can bypass intermediate stages. Together with the work environment, the decision sequence forms a closed loop. Actions change the state of the environment, which is monitored by the decision maker in the next pass through the decision ladder. [Adapted from Rasmussen (1976) with permission from Plenum Publishing Corp.]

systematic way the human-machine interaction in highly automated systems where system designers, operators, and process computers will have to cooperate intimately in the control decision during complex disturbances. The background knowledge and data needed vary considerably for the different phases of the decision sequence. So do the capacity requirements for memory and information processing. Therefore, the designer's prior analysis, the operator's "on-line" decisions, and the support functions of the computer will have different roles during the different phases. Proper allocation of decision tasks should be based on demand-resource matching, considering the requirements of the individual decision subroutines and the capabilities of the designer, operator, and computer. It is therefore important to investigate the conditions under which the individual subroutines of the decision sequence can be formulated generically and separated from the overall decision sequence. In other words, it is important to find out whether the subroutines have some invariant structure across occurrences. Because the identification or diagnosis of system state is a difficult task that, in particular, depends upon the actual conditions, this task is chosen for detailed discussion in the next chapter.

Chapter 3

The Diagnostic Task

In order to formulate this problem in more detail, we will consider the function of state identification in different situations during plant operation with reference to the decision ladder of Figure 2.1. It is important to note that "system state" is used here in a broader sense than that of mathematical state-space representations. It also includes the status of system configuration, such as "valving and switching states," as well as "failed states" of equipment.

The simplest situation is the identification of the states related to those control actions during the normal plant conditions that have been considered during design of the automatic control system. In such cases, analysis has identified the relevant state-action associations, which are then implemented in the control system. At the lowest levels in the control hierarchy, this is done in the form of dedicated feedback loops where deviation between a measured variable and its set-point value leads to compensation through adjustment of a predetermined parameter. At higher-level sequence control, predetermined patterns of plant variables are used as templates to test whether conditions for some planned actions on the plant are satisfied. In such cases, all higher-level functions of the decision ladder have been considered only by the system designer; during on-line decision making, they are bypassed by stereotyped links between a pattern of observations and a set of actions. Similar state identification is used to protect the plant in case of disturbances and faults involving consequences that occur too rapidly or are too drastic in nature to be considered suitable for on-line human decision making. Again, stereotyped links can be implemented in an automatic safety system acting on a predetermined pattern of state variables. An important feature of this state identification mode is that identification takes place with reference to data pattern templates and relates directly to actions specific for each case.

The differentiated control actions necessary in response to faults and

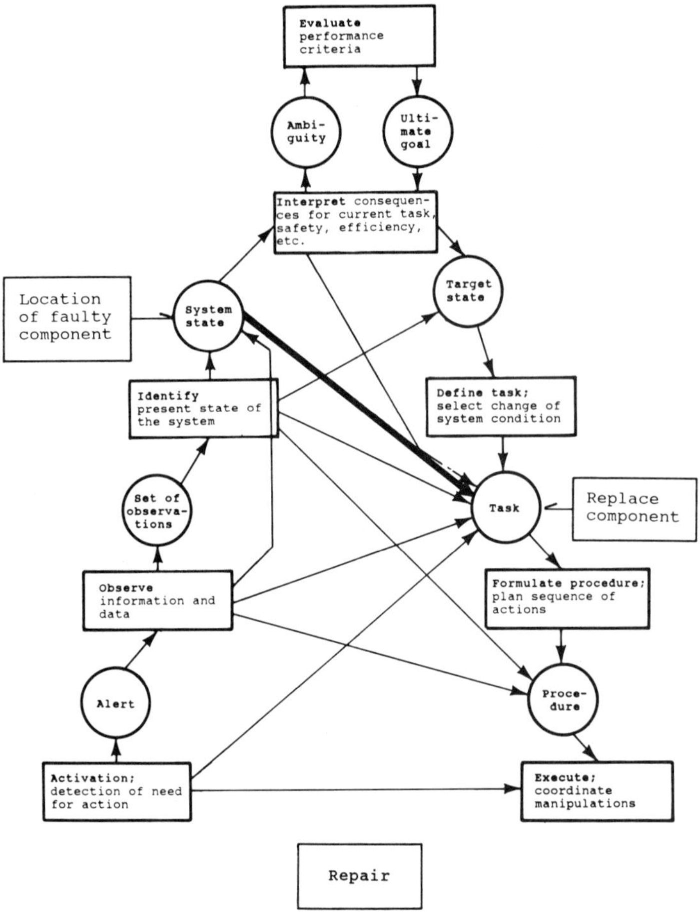

Figure 3.1. The diagnostic phase of a decision sequence can be separated from its context only in certain situations as, for example, in workshop maintenance. In process control, the diagnostic phase will depend on goal priorities and available resources in a circular relationship.

disturbances depend on more flexible identification of individual disturbances, and on-line decision making will include higher-level processes. The extent to which this occurs will, however, depend on the actual circumstances. Consider first the state identification or diagnosis involved in maintenance and repair. By definition, the target state of repair activities is the normal, physical configuration; the fault has to be removed by replacement of the damaged part. This task and the related procedure are well determined, and diagnosis simply means location of the faulty part, i.e., identification with reference to the normal, physical state as illustrated in Figure 3.1.

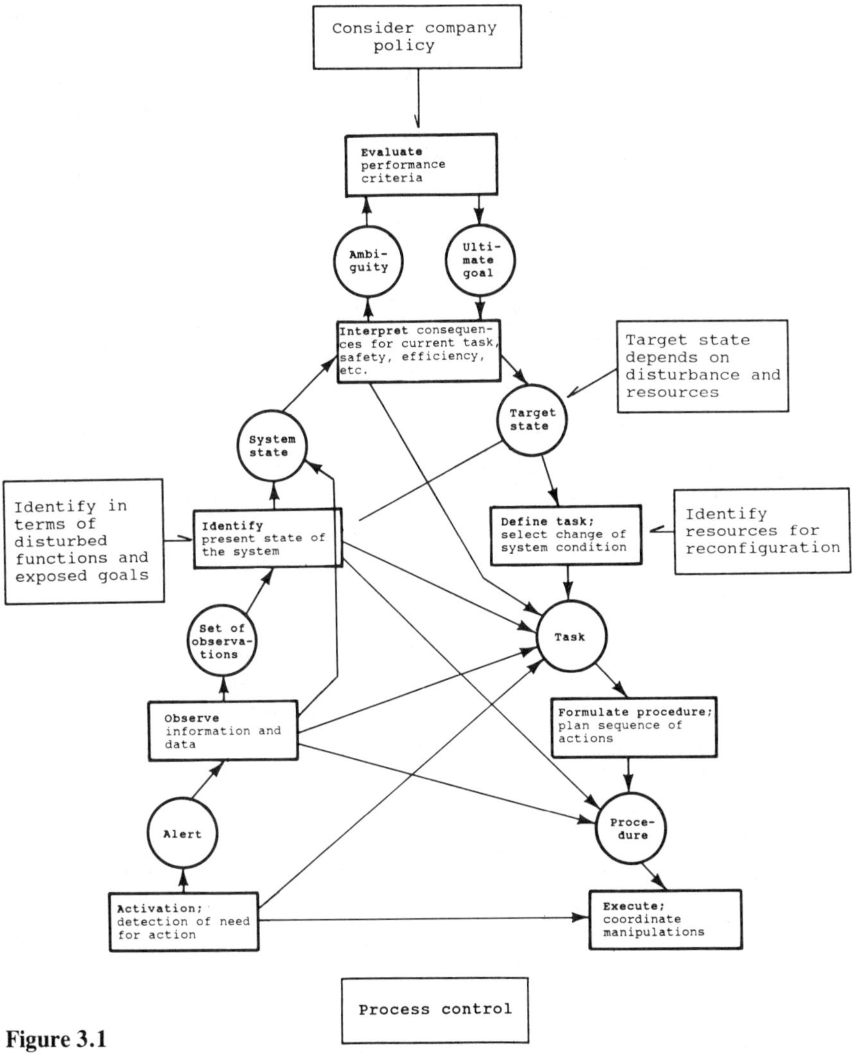

Figure 3.1

However, repair cannot be the first control response to a fault during plant operation. For the initial control response, the target state is not given a priori. Even the goal, relative to which control strategies must be planned, will often be ambiguous–priority judgment must be made whether control actions should be chosen to maintain production, to protect equipment or people, or to minimize downtime in case of plant shutdown. In this situation, it is not possible to separate one well-defined diagnostic task from the overall decision sequence; the content of the task is not simply an identification of the primary cause in terms of equipment faults. Before attention is paid to this problem, the main concern will probably be to see whether the primary mass or energy balances have been disturbed and may lead to major risks, which necessitates a search for deviations from normal in terms of major flow paths in the system. In order to evaluate resources for a compensation of such a disturbance, in terms of heat sinks for improved cooling, for instance, the disturbance must be considered with reference to the actual anatomic configuration of the system and to possible reconfigurations.

In the case of disturbances of an industrial process plant, therefore, the diagnostic task implied in supervisory control may very likely require an iteration between consideration of the functional properties of the system at various levels of abstraction. The implications of the actual plant state and of possible corrective actions must be evaluated against the overall operational goals and constraints; the functional relationships in the actual state should be considered with reference to the normal function; and the possible alternative resources in terms of equipment and supplies that can be used to counteract the disturbance must be reviewed. This process is intrinsically circular inasmuch as the priority ranking of goals depends upon the nature of the disturbance and, at the same time, the level of state identification will depend on the goal being pursued.

From this circularity it follows that an iteration back and forth between the various steps in the ladder will be required in any task context where multiple goals and multiple causes have to be considered. Consequently, the decision ladder is not a model of the decision process of real-time decision making but, instead, a framework representing the informationally logical relationships among states of knowledge and, as such, useful as a map upon which the decision process can be represented (see Chapter 8, Figure 8.4).

The conclusion of this discussion is that in order to generalize a description of the diagnostic task, it is advantageous to consider two aspects separately. The first is the search strategy that is applied in order to locate the disturbance. The second is the problem space, i.e., the description of the physical system, in which the search is performed.

Other phases of the decision process also depend on a description of the system at different levels. Counteracting a fault in the system during operation depends upon a redundancy in the purpose/function/equipment relationships of the system, and an analysis of these relationships is also an important part of planning phases of supervisory decision making. That is, a systematic description of the system to be controlled in terms of multilevel framework, a functional abstraction hierarchy, is an important basis for design of supervisory control systems.

Chapter 4

Abstraction Hierarchy

The description of a man-made physical system in terms of a functional abstraction hierarchy will be discussed in some detail because it appears to be an effective representation of the context in which supervisory decisions are made.

Supervisory control decisions based on functional analysis, i.e., control decisions in unfamiliar situations for which "know-how" short cuts are not feasible, depend on a prediction of the responses of the system to the intended control action. Such predictions are based on knowledge about the functional properties of the system, e.g., knowledge about the relationships among physical variables, about the possible use of components, and about the ways in which events will propagate through the system.

The way in which the functional properties of a system are perceived by a decision maker very much depends upon the goals and intentions of the person. In general, objects in the environment in fact only exist isolated from the background in the mind of a human, and the properties they are allocated depend on the actual intentions. A stone may disappear unrecognized into the general scenery; it may be recognized as a stone, maybe even a geologic specimen; it may be considered an item suitable to scare away a threatening dog; or it may be a useful weight that prevents manuscript sheets from being carried away by the wind–all depending upon the needs or interests of a human subject. Each person has his own world, depending on his immediate needs. A classic discussion of the relativity of the environment depending on the perceiving organism is given from a biologic point of view by von Uexküll and Kriszat (1934).

A structured representation of a physical system as it appears to a supervisory decision maker will be important for a framework suitable to describe human-machine interactions. The decision will be dynamic, and the perceptions of a decision maker will change through time, even within the same task sequence, because several different categories of relationship among concepts will be used for different purposes. The most typical categories that we have met in protocols and interviews are mentioned here to illustrate the context of the functional abstraction hierarchy that will be considered in detail.

Set membership relations of a generic hierarchy will be used to label or name objects in order to answer questions such as "What is this?" This kind of category will be familiar from biologic classification; a technical example could be: process plant component -> pump -> centrifugal pump -> specific type of pump. To discuss the composition of a given piece of equipment, consideration is structured according to *whole-part relationships* in a decomposition hierarchy, such as: diesel generator -> oil supply system -> injection pump -> pump bearing. Variation along the whole-part dimension is necessary for control of the span of attention and continuously changes the definition of "objects" to consider. In order to identify or describe objects or concepts, categories in terms of *descriptive attributes* are used, frequently represented, however, by prototype members. To predict the course of events during interaction with a physical system, *cause-and-effect* relationships are used to predict the propagation of changes through a system. It should be mentioned here that relationships are frequently found that cut across these more formal categories, such as *episodic relationships* that refer to the context of prior experience.

For supervisory control decisions, we have found very useful a description of the functional properties of a system according to goals-means or *means-ends relationships* in a functional abstraction hierarchy. Such a hierarchy describes bottom-up what components and functions can be used for, how they may serve higher level purposes, and, top-down, how purposes can be implemented by functions and components. Various forms of functional abstraction hierarchies including means-ends relationships have typically been used for design activities.

Descriptions of man-made systems in terms of a multilevel abstraction hierarchy are well known in architectural design (Alexander, 1964). They have been extensively used for organizing data bases for computer-aided design and manufacturing (CAD/CAM) (Eastman, 1978). From a philosophic point of view, they have been discussed by Polanyi (1958, 1967).

During system design and supervisory control, the description of a physical system will typically be varied in at least two ways. The description can be varied independently along the abstract-concrete dimension, representing means-end relationships, and the dimension representing whole-parts relationships; the first dimension relates to the concepts used for description; the second relates to the level of detail chosen for description. It will be seen later that changes along the two dimensions are very often made simultaneously, but can in fact be done separately. In the organization of abstraction hierarchies for design, there has traditionally been no major effort to distinguish between aggregation and abstraction. In the present approach, however, this distinction is found to be important and will be maintained throughout the discussion. The resolution of a description is controlled by aggregation-decomposition of the elements used to represent system properties. For example, a system can be considered as a

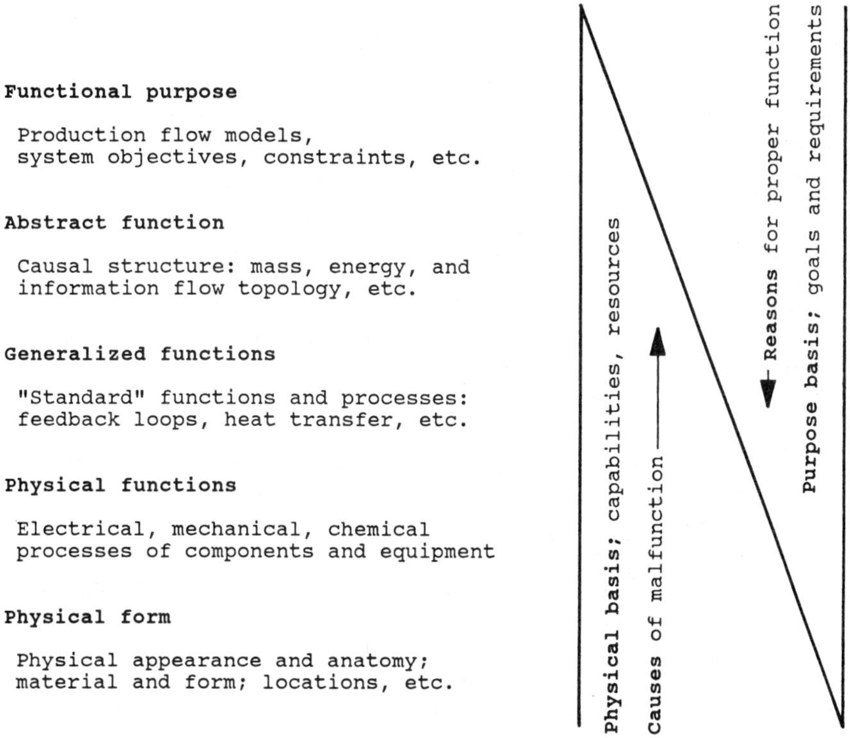

Figure 4.1. A means-ends abstraction hierarchy used for representation of functional properties of a technical system. [Reproduced from Rasmussen (1983) with permission from IEEE.]

hierarchy of parts, ranging from elementary components such as nuts and bolts, through equipment such as pumps and motors, to the complete plant. In the abstract-concrete dimension, the functional properties of the system are represented by concepts that belong to several levels of abstraction, from the physical form of equipment, through processes and functions, to aspects of purpose. The number of levels of abstraction to consider depends on the kind of system in question and the purpose of the description. For representation of the functional properties of a physical system, the levels shown in Figure 4.1 are found to be useful (Rasmussen, 1979b, 1983).

The space between the immediate physical appearance of the system and the functional purpose is bridged by five levels of description. Abstraction, in the present context, does not simply mean the removal of concrete detail as it may in a purely generic hierarchy. When moving from one level of description to the next higher level, information representing the physical implementation is discarded, but at the same time information related to the general cofunction of elements is added which, for man-made systems, means information related to the purpose of the system, i.e., the reason for the actual configuration. The different levels represent the various more or less standardized languages and concepts used by professionals in typical work situations during conceptualization, design, and operation of a system. The number of levels in use may therefore very well depend on the professional field.

The Level of Physical Form

The lowest, most concrete level of abstraction is the representation of physical form, i.e., the physical appearance and configuration of the system and its parts. This level includes descriptions such as pictures, perspective drawings, maps of territories, and locations of equipment as well as scale models. Even at this level, the representation is not neutral or objective. The purpose of the system considered will very clearly control what is selected for representation, how the physical environment is structured in components and parts, and the resolution that has been chosen for description (consider, for instance, city maps intended for various planning and sightseeing purposes). Descriptions at this level are vital for interaction with a system in order to find one's way and to identify parts and components to manipulate. The relationship with higher levels is very often reflected in the labels and names attached to the physical items, reflecting their functions and reason for their presence in the physical map. Labels like pumps, valves, switches, and meters refer to the physical function of the components, and not to their form or appearance. This level is clearly the one of most importance for control of activities during plant construction, in the form of architectural drawings, piping diagrams, and component inventory lists.

The Level of Physical Function

The level of description related to physical function represents the physical (i.e., the mechanical, electrical, or chemical) processes of the system or its parts. At this level, the description of functional properties is tightly related to the physical implements, which will very often be by means of standardized technical components of widespread use. The description is focused on the specific physical equipment and the physical variables used to characterize its functional states. As examples, consider diesel engines with their processes of combustion cycles and mechanical transmissions; transistors and their characteristic electrical relations in the form of signal matrices or graphic data sheets; centrifugal pumps with pump characteristics and limiting properties. This is the level of general professional knowledge about typical components and their properties. The properties of the physical items that are included in the descriptions are very clearly determined by their typical functions, i.e., the reason for their use and their limiting properties. The level of detail, the resolution of the description, depends upon the task and profession of the people interacting with a technical system. For a power plant staff, a diesel electric generator will probably be a physical component, whereas a repair technician will consider an exhaust valve a component. This is one of the reasons for keeping the level of abstraction and the level of decomposition separate in the formal framework.

The level of physical function is the level at which physically limiting properties are represented and at which causes of malfunctions are typically identified. Physical changes in components have functional consequences that propagate up through the levels of abstraction. The level of decomposition at which the fault is defined depends on circumstances. A power plant operator may locate the cause in a pump and from this make his decisions, whereas the repairman has to identify the broken bearing seal.

The Level of Generalized Function

The description at the level of physical function is based on concepts and relations characteristic of the physical process of the related components. At the next higher level of generalized function, the tie to the physical implementation is cut, and the concepts and language used for description are related to functional relationships that are found for a variety of physical configurations and have general, but typical properties. Examples are feedback loops, which have very characteristic properties, independent of their implementation in the form of mechanical, pneumatic, or electrical systems; properties of "cooling functions" can be established independently of the techniques used for pumps and heat exchange; the characteristics of an intermediate frequency amplifier of a radio receiver can be stated in terms of gain and selectivity curves without reflecting the design in terms of tuned circuits or ceramic vibrators. Typically, the generalized functions will represent the cofunction of equipment or components based on different physical functions, and represent the functional structure of a system at a level of decomposition that is above the level of standard components. There is normally no one-to-one relationship between the physical function of a component and its potential role in a generic function. In general, there is a potential many-to-many mapping that represents the functional redundancy that is necessary for the freedom of operators to compensate for the effect of disturbances. Braking your car generally depends on the physical function of the brakes; if, however, they fade going down a mountain road, you try to brake by means of putting the engine in low gear.

There is usually a clear distinction between the properties of a system that are represented at the levels of physical and of generalized functions. Descriptions of physical functions are oriented toward the functioning of physical components and equipment, i.e., models are structured according to available components. Descriptions at the generalized level deal with functional relationships that are widely found independent of material manifestations. This means that generalized functions are structured according to available models of functional relationships, i.e., the tools for analysis.

Generalized functions are typically the subjects of theoretical studies in classical branches of engineering, and descriptions are based on laws of nature, characteristic of the various professions such as Newton's laws in mechanics, Kirchoff's laws in electrical engineering, etc., and their specific practical derivations. Generalized functions are, for instance, power supply, heat transfer, feedback control, and nuclear fission. Descriptions at the lower level of physical function are typically found in equipment manuals, component specification sheets, and textbooks for technicians and equipment designers. Descriptions at the level of physical functions clearly reflect the purpose of a system and the reason for the presence of the physical components. Description of a physical item like a pump presupposes pipes and fluids and therefore its purpose. In the same way, descriptions of generalized functions reflect the cofunction of various physical processes and, therefore, the reasons behind system design.

At the levels discussed so far, the functional properties of a system are represented in terms of the interaction among a set of typical processes, functions, or physical parts. To represent the overall function of a particular system or type of system by a consistent model, it may be necessary to move up in abstraction to a level that is independent of the actual physical and functional properties, but more tightly related to the intended, proper functional state of the system and its coupling to the environment.

The Level of Abstract Function

At the level of abstract function, the overall function of a system can be represented by a generalized causal network, e.g., in terms of information, energy, or mass flow structures reflecting the intended operational state or, more specifically, in terms of flow of products, commodities, or monetary value through the system. Here we are in the domain of information theory and the domain of laws of conservation of energy and mass. The laws and symbols at this level form a consistent structure that is device- and process-independent and satisfied by the system design. In his discussion of substance versus function, Cassirer (1923) characterizes the concept of energy as follows: "Energy is able to institute an order among the totality of phenomena, because it itself is on the plane with none of them; because lacking concrete existence, energy only expresses a pure relation of mutual dependency." Correspondingly, a consistent formal model of the functional properties of a system in device- and function-independent terms can be developed in the form of a mass and energy flow map. Such a consistent model is very well suited for identification of control requirements and for automatic diagnosis (Lind, 1981, 1982).

At this level of overall functioning of a system, a model only has meaning when considering a working system, in that the actual "mutual dependency" is determined by the cofunctioning of all the elements of the system, and upon the actual couplings to the environment such as inputs and power supplies. Therefore, representation at this level depends on knowledge about the purpose of the entire system and about the reasons for the actual structure.

The transition from the level of generalized function to that of abstract function is probably most evident when considering information-processing systems. Here, a set of coding conventions relates the actual functioning of the system at the physical and generalized levels to the abstract function in terms of information processes. The abstract function represents the semantic content of the physical signals and, hence, the overall organizing principle. Consider, for instance, the relation between the logic functions of an information process and the underlying generalized functions of a digital system, a relation that is based on a set of coding conventions. Also for energy producing or converting systems, the overall function of such systems is conveniently described at this level of abstraction in the form of the intended patterns of energy flows.

Abstract representation of organizing principles for complex systems is also well known from the natural sciences for which they act as a kind of "reason" from which system properties can be derived: the first and second law of thermodynamics, the "survival value" of Darwin's theory, the principles of "least work" of Hamilton's theory, etc.

The Level of System Purpose

At the highest level of abstraction, the purpose (i.e., the intended functional effect of the system upon its environment) is described. This can be done in terms of simple, quantitative input-output specifications or in terms of the environmental, functional relationships of which the system is a part. Very often, however, the operational state is not totally specified and the ambiguities in the possible states of operation have to be determined by the control system by optimization from general criteria related to, for example, economy and conservation of resources.

For many kinds of systems, the ultimate purpose is related to the lower level of abstraction: for instance, the purpose of mechanical manufacturing plants and of treatment plants, such as washing machines, will clearly be related to the level of physical function. Even there, however, the use of higher-level descriptions may be necessary to be able to consider system organization and performance, for instance, with respect to energy consumption and resource conservation. Examples of the functional properties of specific systems related to the various levels of abstraction are shown in Table 4.1.

Use of the Abstraction Hierarchy

At the lower levels of functional abstraction, elements in the description match the component configuration of the physical implementation. When moving from one level to the next higher level, the change in system properties represented is not merely removal of details of information on the physical or material properties. More fundamentally, information is added on higher-level principles governing the cofunction of the various elements identified at the lower level. In man-made systems, these higher level principles are naturally

Table 4.1. Examples of Descriptions in the Means-Ends Abstraction Hierarchy.

Washing machine	Manufacturing plant	Computer system
Purpose		
Washing specifications Energy waste requirements	Market relations Supply sources Energy and waste constraints Safety requirements	Decision flow graphs in problem terms
Abstract function		
Energy, water, and detergent flow topology	Flow of energy and mass, products, monetary values Mass, energy balances Information flow structure in system and organization	Information flow Operations in boolean logic terms, truth tables Symbolic algebraic functions and operations
Generic function		
Washing, draining, drying Heating, temperature control	Production, assembly, maintenance Heat removal, combustion, power supply Feedback loops	Memories and registers Amplification, analog integration and summation Feedback loops, power supply
Physical function		
Mechanical drum drive Pump and valve function Electrical/gas heating circuit	Physical functioning of equipment and machinery Equipment specifications and characteristics Office and workshop activities	Electrical function of circuitry Mechanical function of input–output equipment
Physical form		
Configuration and weight, size "Style" and color	Form, weight, color of parts and components Their location and anatomical relation Building layout and appearance	Physical anatomy Form and location of components

derived from the purpose of the system (i.e., from the reasons for the configurations at the level considered). Change of level of abstraction involves a shift in concepts and structure for representation as well as a change in information suitable to characterize the state of the function or operation at the various levels of abstraction. Thus, a decision maker will ask different questions regarding the state of a physical system, depending upon the level of abstraction that is most suited to formulate the actual control task.

At low levels of abstraction, models are related to a specific physical world that serves several purposes. Models at higher levels of abstraction are closely related to a specific purpose that can be met by several physical arrangements. The functional abstraction hierarchy is therefore useful for a systematic representation of the many-to-many mapping in the purpose/function/equipment relationship, which is the context of supervisory decision making. For a process at any level of the hierarchy, information on proper function is obtained from the level above, and information about available resources and their limitations is obtained from the level below (Rasmussen and Lind, 1981).

In the present context, an important use of the abstraction hierarchy is as a framework for description of the control tasks required to maintain satisfactory system operation. States can only be defined as errors or faults with reference to the intended functional purpose. Causes of improper functions depend upon changes in the physical or material world. Thus, they are explained "bottom-up" in the levels of abstraction. In contrast, reasons for proper function are derived "top-down" from the functional purpose. This distinction is illustrated in Figure 4.1.

During system operation, the task of the control system (which includes the automatic control system as well as human operators) will be, by proper actions on the system, to ensure that the actual state of the system matches the target state specified by the intended mode of operation. This task can be formulated at any of the different levels of abstraction. During system start-up, for instance, the task moves bottom-up through the hierarchy. In order to have an orderly synthesis of the overall system function during start-up, it is necessary to establish a number of autonomous functional units at one level before they can be connected to one unit at the higher level. This definition of autonomous functional units at several levels is likewise important for orderly breakdown of system functions for shutdown and for reconfiguration during periods of malfunction. For such considerations there will typically be a tight coupling between the concrete-abstract and the part-whole dimension.

During emergency and major disturbances, an important control decision is to set up priorities by selecting the level of abstraction at which the task should be initially considered. In general, the highest priority will be related to the highest level of abstraction. First, judge overall consequences of the disturbances for the system function and safety in order to see whether the mode of operation should be switched to a safer state (e.g., standby or emergency shutdown). Next, consider whether the situation can be counteracted by reconfiguration to use alternative functions and resources. This is a judgment at a lower level of function and equipment. Finally, the root cause of the disturbance is sought to determine how it can be corrected. This involves a search at the level of physical functioning of parts and components. Generally, this search for the physical disturbance is of lowest priority (in aviation, keep flying–don't look for the lost light bulb!).

When a disturbance has been identified and the control task located at a level of

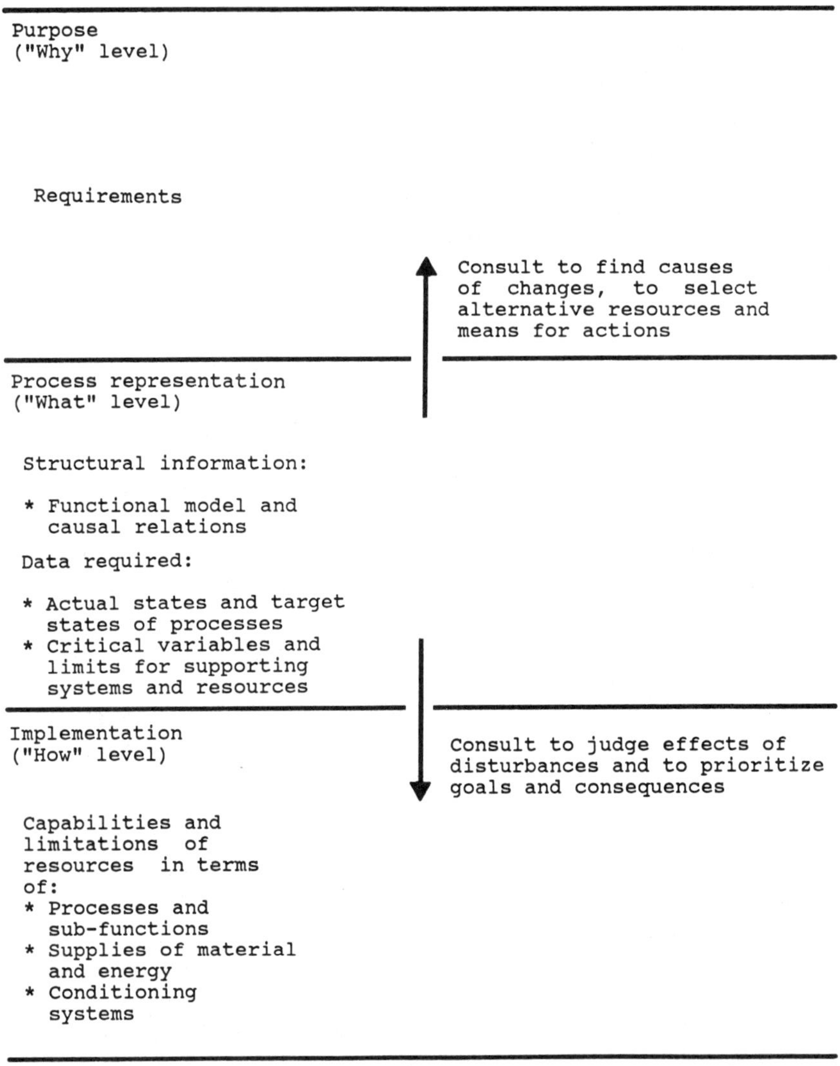

Figure 4.2. Schematic illustration of the relationships among the three levels of abstraction to be considered for a control task.

abstraction as required by the situation, the supervisory control task includes the determination of the target state derived top-down for the chosen operating mode. In addition, the available resources for reconfiguration and limits of capabilities must be derived from levels below.

A decision task in a particular situation can, as has been discussed, be formulated with reference to a process at any level of the abstraction hierarchy. For the task, the operator will typically need information from the level

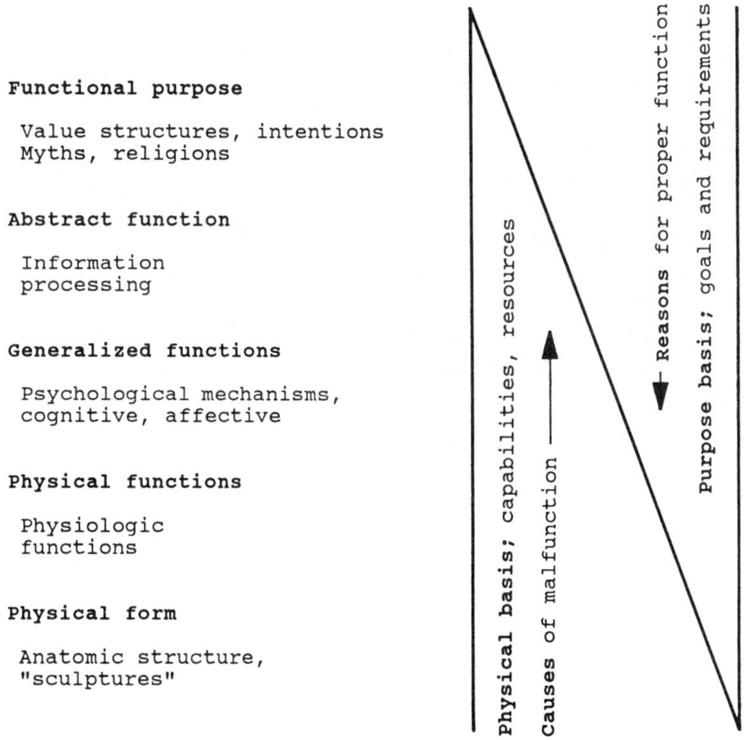

Figure 4.3. The abstraction hierarchy used for description of human functional properties. [Reproduced from Rasmussen (1983) with permission from IEEE.]

considered, representing the functional structure and state of this process, i.e., regarding what is controlled. But the operator will also need information from the level above, which is related to the immediate purpose of the control decision–i.e., *why* it is made—as well as from the level below–i.e., *how* a decision can be implemented (see Figure 4.2). In a particular situation, a decision maker will not be aware of the multilevel structure of the representation he is using of the entire system. In making a decision, he will only have a certain part of the system within his span of attention. To the decision maker, this part is "the system" to be represented in the actual situation, and the rest of the system is part of the "environment." In considering only this more restricted "system" in a decision, only the three levels of what (process), why (purpose), and how (implementation) will be involved. However, when the focus changes in accordance with the requirements of changing work situations, these three levels of abstraction will generate the full abstraction hierarchy of Figure 4.1 by recursion as the attention shifts, as it was discussed for diagnosis during disturbance analysis.

Frequently, other persons will be part of the environment with which a particular person interacts, and for which he or she has to use mental models in order to cope with unfamiliar situations. In order to prepare for the discussion of models of humans, it should only be noted here that various levels of abstraction can also be used to model human "functional" properties, and an analogy of the levels discussed in Figure 4.1 is drawn for "models of man" in Figure 4.3. A more detailed discussion can be found in Chapter 8.

In conclusion, the functional abstraction hierarchy representing means-end relationships serves the purpose of a systematic description of the context in which supervisory decisions are made. The other basic aspect to be considered is the information processing strategies humans can use for the different phases of the decision sequence and the performance criteria that control their choice in the actual situation. In the next chapter, these aspects of a diagnostic task are discussed from analysis of "real-life" performance, based on verbal protocols.

Chapter 5

Analysis of Strategies for Diagnosis and Performance Criteria in Real-Life Tasks

In this chapter, the information process related to diagnosis will be considered. The most complex data processing takes place when diagnosis is required during disturbances of system operation. Today, this task depends on the performance of human operators, but the trend is to use the data-processing capacity of computers to support the task. To reach a proper human-computer cooperation, it will therefore be necessary to study the diagnostic strategies that are actually used by operators in different situations. From this result, one can generalize and formulate a set of formal strategies that can be used as a basis for system design. Furthermore, it is very important to analyze the subjective preferences and performance criteria that guide an operator's choice of strategy in a specific situation. Unless these criteria are known, it will not be possible to predict the strategy that an operator will choose, faced with a specific interface design.

This implies the study of mental procedures used during real-life work conditions for which the use of interviews and verbal protocols are suitable tools. During the high season of behaviorism in psychology, there were serious objections to the use of introspection and verbal statements. However, use of verbal protocols to identify data-processing strategies during performance is not based on introspection; one is not asking the person to turn his attention away from the actual tasks toward his internal processes. One is just asking him to express his intentions, thoughts, and needs (i.e., to externalize the internal verbalization that he may use anyway).

The quality of the protocols that one can obtain depends very much on the work conditions and the nature of the task. The diagnostic task aimed at repair has a much more well-defined nature than the diagnostic phase of supervisory control (see Chapter 3). Furthermore, it is very difficult to obtain a reasonable number of good protocols from diagnosis in real-life plant operation due to the stochastic and infrequent occurrence of the related events. It was therefore decided to perform the basic study of diagnostic strategies in an electronic instrument repair shop (Rasmussen and Jensen, 1973, 1974). The usefulness of the generalized results was later confirmed by an analysis of protocols from diagnosis in computer systems and process plants, together with an analysis of error reports from power plants.

The situation in the electronic maintenance group of the Risö National Laboratory was deemed to be very suitable because there was a close personal contact and working relationship between the people conducting the experiments and the repairmen, and there was the natural interest of the maintenance group of a scientific institute to be directly involved in a research program related to its own professional methodology.

An Example

A discussion of the results of one analysis from a simplified example based upon the main features of one of the actual cases will be used to illustrate the approach. The case considered is one where a digital scaler displays two digits simultaneously in one decade, but otherwise functions normally. Now the task is to obtain a reference to the location of the faulty component from the response of the system and by appropriate measurements.

In the present case, there is a close relation between the faulty parameter of the system response (i.e., the fault in display of the second decade) and a well-defined part of the system. The technician's interest will then be quite naturally limited to the circuitry connected with the second decade. Further reference to the location of the fault may now be obtained in different ways. It may be based upon detailed observations of the actual faulty response and consideration of the internal anatomy and functioning of the system. In the case under consideration, a design engineer localized the fault to a specific resistor in the decoder directly from the response, using his knowledge of the digital code and a diagram of the circuitry of the decoder. This method can, of course, also be chosen by a trained maintenance technician and, judging from textbooks for the training of such personnel, some authors consider it to be the "intelligent" method, i.e., to take a few, carefully chosen measurements and use the observations in careful reasoning based upon functional understanding of the system.

Our observations indicate that a trained technician is most likely to choose another method, i.e., that of scanning through the faulty decade by a rapid sequence of good/bad checks of the actual signals against normal signals, which are measured in one of the other decades, or which are found on the circuit diagram. In this way, the fault may rapidly be localized to the decoding circuit. In the circuit diagram, this circuit is seen to contain less than half a dozen resistors. Therefore, rather than evaluate their function, it may be preferred to scan through the resistors by good/bad checks with an ohmmeter. Thus, an open circuit resistor is found.

General Features of Diagnostic Performance

Although very simplified, this example illustrates some of the general features of the procedures found in the data collected:

1. The basic feature of routines used may vary greatly in several respects. In the example given above, the designer used only a few observations, but employed complex data processing in his decision procedure. His procedure is related to the anatomy and internal functioning of the specific system and to the actual faulty condition. He treats several observations simultaneously, and his procedure is informationally economic.

   The trained technician uses many observations in a sequence of simple decisions. His method is a general search procedure that is not dependent upon the actual system or specific fault. He treats the observations individually in a stream of good/bad judgment that is informationally uneconomic, but fast.

2. The technician defines his task primarily as a search to find where the faulty component is located in the system. He does not consider it to be a problem-solving task, which involves explaining why the system has the observed faulty response and understanding the actual functioning of the failed system.

3. The procedures are organized as a search through a system that is viewed as a hierarchy of units. The system is composed of a number of subsystems: amplifiers, scalers, deflection generators, etc. Each subsystem has easily identifiable units such as amplifier stages, flip-flops, and oscillators, and these units have components, e.g., transistors, capacitors, and resistors.

4. The general structure of the search can be broken down into a sequence of search routines, which are used to identify the appropriate subsystem, stage, or component. The structure illustrates how the technician attempts to limit sequentially his current field of attention. He is constantly asking the question of where to look next and thus tries to extract topographic references from his observations.

5. A topographic reference from the observations is typically obtained in three different ways depending on very different depths in the consideration of the internal anatomy and functioning of the system.

Analysis of the protocols identified a number of such diagnostic subroutines that can be distinguished by the type of information that refers the technician toward the location of the fault. The strategies are discussed in detail in the following sections.

The Functional Search

In the functional search, the topographic reference is obtained from the normal functional relationship between a feature in the system response and a specific part of the system. A good example is troubleshooting in a TV receiver. The repairman will scan the features of the picture in a stream of good/bad judgments and turn his interest to the subsystem related to the faulty feature. If the picture is too low, he will perform a search in the vertical deflection generator. The functional search is quite naturally the opening move in complex systems having subsystems with specific functions that are individually recognizable in the overall system response. However, the routine may also be used later in the procedure when the technician is faced with more complex data patterns such as wave forms on oscilloscopes.

The search is a special type of the topographic search discussed in the next section. The information pattern is scanned and familiar features are judged individually in a stream of good/bad judgments, and only results of judgments are normally used to control the next activity. If a response feature is judged faulty, attention is typically turned immediately toward the subsystem related to that function, and a routine search is then performed in the subsystem. Information related to the observed mode of failure of the function is used in a very cursory manner.

In not a few cases, the technician studies the faulty response in great detail, but the information is not used to any large extent to control the search that follows. Sometimes the information is recalled later in the record to confirm a hypothesis found by routine search, and perhaps accompanied by a remark indicating a "eureka feeling" when the hypothesis appears.

Faced with a multiparameter pattern of information, the technician normally has a good opportunity to deduce not only rather precisely what is the cause of the faulty response and where the fault is to be found, but also what sort of search procedure will be the most efficient. However, one of the clear indications found in these experiments is that information available is not used efficiently in that way, even when faults were simulated to invite the use of shortcut methods by functional reasoning and evaluation.

The Topographic Search

In the topographic search, reference to the location of the fault is obtained from the topographic location of a measuring point. The system is scanned by a sequence of measurements, and the observations are subject to simple, individual good/bad judgments.

The search is normally a test of performance along a main signal path in the relevant subsystem. The circuitry along the route is seen as a row of familiar units (e.g., amplifier stages), and by a sequence of rapid judgments the stage is localized in which the signal disappears or a faulty signal appears.

When turning to a subsystem to perform a topographic search, normally very little information from previous observations (e.g., the nature of the fault) is carried over to assist in planning the search. The choice of the parameter to be used in the search is very dependent upon those norms for judgments that are immediately available. In some cases, the parameters chosen for the search cannot lead to the location of the fault, and information clearly indicating this may actually have been recorded by the troubleshooter prior to the decision to start the search. However, the decision about where to look is often the only connection with the previous search. This may seem very inefficient, but still it should be remembered that a simple search of typically 5-10 measurements in a very rapid sequence may very often prove successful. Therefore, in the long run it may pay to take a chance and not consider every decision carefully.

In the records collected, it was found that the selection of the route and the steps of the search sequence are generally based upon the wiring diagram in a manual of the system. For this application, the diagram is not viewed as a functional description of the system, but solely as a topographic map, showing the information highways. Measuring points are chosen along the route of search at locations where convenient norms for judgments are available (e.g., where the diagram gives good reference data such as bias voltages or signal wave forms). The search is performed as a rapid scan along the route chosen and no judgments as to whether some of the steps will be informationally redundant seem to be made.

Literally speaking, all data collected during the search sequence are immediately judged good or bad individually. For these judgments, the technician needs some model or description of the normal state of the system that supplies him with reference norms for each observation. The presence of a set of convenient reference norms is often directly stated in the record as the reason for the choice of the route and parameter used for the search.

If reference standards for the judgments are not otherwise available, the technician has to work them out himself by deduction from his understanding of the normal function of the circuitry, a task that is normally supported by the wiring diagram. If he also has to plan the route for search from a functional understanding of the circuit, he has to maintain mental models at two different functional levels of the abstraction hierarchy simultaneously, which is a considerable task. A model at one level is necessary to control the route of search; this model has to be related to the signal or information flow and thus to the function of the entire subsystem at the level of abstract function. Models at the other level of physical function are needed to supply reference norms for the individual judgments and thus have to be related to the detailed functioning of the subunits along the search route. In that case, the procedure is slow and hesitating, probably due to the considerable difficulty in maintaining models at different levels simultaneously.

If the necessary reference data for the judgments are hard to find, the measuring point or even the route may be discarded. Switching of attention to another route or field of search is often preferred to the effort needed to establish reference norms by functional reasoning.

Search by Evaluation

Search by evaluation of the fault is used when the technician derives the topographic reference from the actual faulty response. This derivation implies an analysis of the information observed with respect to the specific instrument and its actual state of operation. The progression in this transformation may be illustrated as:

1. Observation (i.e., data observed describing the failed state).
2. Cause (i.e., what is changed in internal signals or functions).
3. Location (i.e., where the faulty component is that results in the specific malfunction).

To be able to make such transformations, the troubleshooter has to use mental models of the system, relating changes in internal signals, parts, or components to the changes observed in system response. Clearly, such transformations will be much more varied in their individual appearance than are the routine search procedures. Also, the complexity of the transformations varies greatly from rapid statements based upon recognition from previous cases to more complex deductions based upon several parameters and careful consideration of the internal system functioning and anatomy.

Such statements are generally expressed as recognition. The transformations based upon conscious reasoning related to internal functioning of the system are complex and difficult to keep pace with during verbalization, and the records indicate only the surface of the activity. This fact, combined with the low number of cases, allows only very general and subjective attempts at classifying behavior, but at least two groups of procedures seem to be used.

In one type, the technician seems to use a hypothesis-and-test search. He is working from inside the system outwards to the response in a way that could be illustrated as: (a) Establish, by examination of diagrams or by memorizing, a mental model of the normal system anatomy, its signals, and functioning. (b) Then make a guess as to which signal or components might be involved in the faulty response. (c) Modify the model accordingly and (d) evaluate the resulting response pattern. (e) Compare with the data observed to judge the relevance of the guess. This sort of procedure is most clearly expressed when the hypothesis is not a guess, but when the fault is found by another search procedure and the result is tested against system response by functional reasoning.

In other cases, the troubleshooter is working from response data into the interior of the system. The procedure looks like a mental topographic search: from the response pattern and an understanding of the system, the absence of a normal signal or system state along a chosen search route is deduced by functional reasoning. The main difference from the normal topographic search is that the data, which are subject to individual judgment, are not measured directly, but deduced from the system response.

Generality of Procedures and Depth of Supporting System Knowledge

The mental data handling necessary in a cooperative human-machine endeavor implies that the human has available some sort of mental model of the system and a strategy to use this model to process observed data. The mental model as well as the strategy may be supported by external means such as diagrams, drawings, instructions, and rules.

The previous discussion indicates that the mental strategies used in troubleshooting vary greatly with respect to the depth of system knowledge needed as support. At one extreme, technicians have strategies that are based upon only very general professional training and experience; at the other, they have strategies that call for very detailed knowledge of the specific system and the laws controlling its internal functioning. The experiment discussed here demonstrates the technicians' great ability to get around their search problem by means of a sequence of general routines mostly depending upon their general professional experience and background.

In the preferred version of the topographic search, the search procedure consisting of a sequence of good/bad judgments is of very general applicability. The model of the system used for the search has only to supply an appropriate route for search and reference data for the technician's judgments. If a circuit diagram is available that clearly indicates the main signal path and gives sufficient data for normal bias voltages and signals, the mental model needed has only to support the topographic correlation between the diagram and the system. The model need only be based upon professional experience with the visual appearance of typical components and circuits and with the normal layout of the circuitry.

The records also show a clear ability to base a functional search upon very general mental models of the system. The technician will scan familiar features in the response by a sequence of good/bad judgments. If a faulty response feature is found, a general "block-diagram understanding" of the system can refer to the related subsystem. If this is a topographically well-defined part of the system, attention will immediately switch to this system in order to perform a routine search. Even when such faults were simulated so that the system response, as judged by the planners of the experiments, clearly indicated possible shortcut methods if the internal functioning of the system was considered, the technicians normally used their general search routines.

Also, in the search by evaluation of the actual fault mode, the records show a pronounced preference for the use of transformation models that are not closely related to the specific system, but based upon general experience. Thus, an interesting feature of the procedures found is the pronounced ability demonstrated by the technicians to produce overall strategies based on general search routines that are not closely related to the specific instrument. Scanning a high number of observations by simple procedures is clearly preferred to the preparation of specific procedures worked out by studying or memorizing the internal functioning of the system.

Redundant Observations, Impulsive Decisions, and Mental Load

Other fundamental aspects of the procedures quite naturally follow the preference for general methods not closely related to the specific system or its actual fault. A general procedure cannot, of course, be based upon very detailed information found in the observations or measurements. In particular, general methods cannot take advantage of information contained in specific relationships between several observations. This is clearly indicated by other features of the procedures found in the protocols.

The functional and the topographic searches appear as functional or topographic good/bad mappings of the system. Practically speaking, all observations are immediately judged to be good or bad, and only the results of the judgments normally control the next activity. In some cases, the parameters chosen for a topographic search cannot locate the fault, and information clearly indicating this may be mentioned by the technician prior to his decision to turn to that particular search. Often, only the decision about where to look connects the routine to the previous search. During the search routines, no attention is paid to whether a measuring point will be informationally redundant or not.

The dependence of the procedures upon individual observations and judgments corresponds to a general tendency found in the protocols. Instead of making overall plans for the search, the tendency is to use rapid or impulsive decisions all along the search, based only upon the information observed at the moment. This gives, of course, a very individual pattern to the different overall procedures found in the different cases. A main rule for the structuring of the procedures seems to be to follow "the way of least resistance." As soon as an indication is found that a familiar, general search routine may be applied, this is chosen without considering a possible, more efficient ad hoc procedure. There seems to be a "point of no return" in the attention of the troubleshooter the moment he makes such a decision, as discussed by Barlett (1958). Although more information indicating possible shortcut methods or important hints for the next search is clearly available from the observation and is mentioned by the technician, the decision prevents any influence from such information; hence, the next search is a routine, starting from scratch.

The basic difference in the amount of data needed for the different search strategies, and in the complexity of the mental data handling task they impose upon the technician, constitutes important features of the various strategies available to the techncian. There is a complementary relation between these aspects of the routines. The very general procedures are based upon a stream of good/bad judgments, and call for a large number of observations that are treated individually and then left behind. The system is mapped in a rather systematic way by such judgments, and this seems to be a convenient way of remembering the results of past activities. A general impression is that during his search the technician is well aware of his previous judgments. However, the originally observed data are discarded without subsequent recall, although in some cases they seem to build up–unconsciously–a sort of "feeling"; later in the procedure this feeling can initiate hyotheses that appear as "good ideas." The very specific procedures based upon system anatomy and functioning require only a few observations, but the information handling is complex, and simultaneous treatment of several observations and a considerable carry-over of information may be needed in the short-term memory between the individual steps of the procedure.

This discussion focuses the attention upon the mental load on the technician during the task. As discussed thoroughly by Bruner et al. (1966), the mental strategies chosen by the technician may be strongly influenced by the constraints he meets in his limited capacity for short-term memory and inference. The multiple task nature of troubleshooting may make this an important constraint. On a time-sharing basis, the technician has to formulate the route for search through the system by use of a diagram or by reasoning, to locate the route in the real system, to manipulate measuring devices, to establish norms for his judgments from diagrams, experience, or functional reasoning, and to keep track of his overall search.

Several indications of high cognitive strain are found in the data. A good example is a topographic search in a digital system performing logic operations, in cases where the technician has to plan the route and produce reference standards for judgments by deduction from an understanding of the functioning of the circuitry. As discussed earlier, he then simultaneously has to maintain mental models at different functional levels, and this is a considerable task. In this case, the procedure becomes slow and hesitating, and in observation the person seems to be very insensitive to hints that would normally be familiar to him.

The records give several indications that difficulties in one of the subtasks tend to cause simpler procedures to be used in others. This should be taken into careful consideration when generalizing from clearcut laboratory experiments with special equipment that eliminates all secondary subtasks.

Fixations in Routine Search Procedures

The records indicate that technicians have a great deal of confidence that the general search routines will ultimately lead them to the fault. If, for instance, a topographic search turns out to be unsuccessful and fails to result in a local search (which occurs in more than half of the attempts), the preferred decision is to repeat the search by another parameter. If this search proves unsuccessful too, there is a pronounced tendency to return to a search performed earlier. This seems, however, to be a repetition with careful judgments of the observations rather than a more careful evaluation of the actual faulty function. It should be stated that this is not necessarily due to inadequate ability to use functional reasoning, but more likely to the fact that these methods are inherently attractive inasmuch as they consist of fast sequences, which are normally successful in the end. The behavior may be compared with that of most car drivers who prefer, when moving around in a big city, to drive along familiar main streets rather than preparing individual shortcut routes by means of a city map.

If the general search routines ultimately turn out to be unsuccessful in special cases, the technician often seems to "be in trouble." When in trouble, there is a tendency to rely on "good ideas" that admittedly seem to appear in most cases after a break or a period of confusion that serve to break fixations. Such good ideas are difficult to trace. Sometimes, the technician returns to deviations met in a previous search, but passed over without further consideration; sometimes he expresses "a feeling that something is wrong around here"–a feeling that has grown from slight indications during earlier search sequences. In some cases, important information has been recorded several times during routine search without triggering his attention until a period of confusion sets in.

When in trouble, there seems to be no tendency to consider it worthwhile studying the functioning of the circuitry in greater detail by means of manuals or diagrams. During a discussion of the strategies found in the experiment, the technicians stated that as a rule they found the general search routines successful. Apart from the need to obtain "block-diagram understanding" of the system, it was not considered worthwhile studying the internal functioning of the circuitry. "If you run into trouble, better take a break, wait until the next day, or discuss the problem with a colleague." Asked if they could suggest types of cases for which they would find it worthwhile studying the internal functioning of the system in detail, the technicians said that it would be the case if measurements of manipulations could have serious consequences, as in live warning systems–"when a siren is at the end of the wire" or if the working conditions on site are unpleasant, e.g., owing to foul odors as in chemical plants. In other words, it would be the case if "costs" related to observation were high. A later test case in connection with the level control system in a radioactive waste tank system in fact resulted in a very "rational" procedure based upon a careful functional evaluation of the system response in advance and very few measurements on site.

It was also suggested to the technicians that they should use functional reasoning when in trouble in the normal repair shop environment. This did not, however, cause any significant changes in the procedures used in the records made thereafter. The procedures seem to be so highly ingrained that they are difficult to change by suggestion of better procedures. When a troubleshooting task is running, a skilled technician seems to be completely absorbed in the task, and he does not "remember" the suggestion when difficulties arise. The test case with the waste tank system may indicate that, to change a procedure, the technician has to perceive the task in advance as one calling for a special search strategy.

Subjective Formulation of Task and Performance Criteria

Important aspects to consider are the subjective formulation of the task and the performance criteria in the choice among the various search strategies available. This aspect is especially important because the formulation of the trained technicians may be basically different from that of a design engineer, who is, after all, very often responsible for preparing the working conditions for technicians in the form of layouts of systems, instructions, and operating manuals and diagrams.

The experiment discussed here clearly indicates that the task is defined by the technicians primarily as a search to find where the fault originates in the system. They are faced with a system that they suppose has been working properly, and they are searching for the location of the discrepancy between normal and defective states. They do not see the task as a more general problem solving task in order to understand the actual functioning of the failed system and thus to explain why the system has the observed faulty response. On the other hand, in his normal work in a development laboratory, a design engineer does not think in terms of standards for normal operation, but considers it his task to understand the basic functioning of the system and to test observations made during his experiments against his conceptual intentions.

Are the procedures found in the protocols rational? What is rational depends upon the performance criteria adopted by the person. Normally, a reasonable criterion for a maintenance technician is to locate the fault as quickly as possible, and only in special circumstances will his criterion be that of minimizing the number of measurements as discussed above. From this point of view, the procedures found in our records are rational because in most cases the faults were found within very reasonable time. (A good demonstration of the difference in troubleshooting time involved in the normal routine cases and calling for more elaborate methods is given in Wohl's (1981) analysis of time distribution in field repair data.) With his theoretical background, the system designer may quite naturally value as rational the "elegant" deductive procedure, which is informationally very efficient and based upon a few observations, but this criterion is not the appropriate one on the basis of which performances in real-life maintenance work can be judged.

In the present context of supervisory control, this analysis of troubleshooting strategies serves to identify several diagnostic strategies that have very different requirements with respect to information and processing resources. It also illustrates the kind of subjective performance criteria that may control the choice of strategy in the actual situation.

Chapter 6

Generalized Strategies for State Identification and Diagnosis

Based on the results of this study of mental strategies in real-life diagnosis, it is now possible to generalize, and to propose a set of possible strategies that can be used for design of operator support systems in process plant control. In this section, the typology of such a set of possible strategies will be developed (Rasmussen, 1981).

As discussed in the previous chapter, the object of the diagnostic search in a supervisory control task may vary: to protect the plant, the search may concentrate on patterns of critical variables related to stereotyped safety actions; to compensate for the effects of the change, search for alternative functional paths bypassing the effect of the change will be appropriate; to restore the normal state, search in terms of the initiating physical change is necessary. In consequence, the diagnostic task implies a complex mental process that very much depends on details of the actual situation as well as the operators' skill and subjective preferences.

Fortunately, to support systems design, it is not necessary to have detailed process models of the mental activities that *are used* by the operators. System design can be based upon higher-level structural models of the mental activities that the operators *can use* and their characteristics with respect to human limitations and preferences. If such a design is successful, operators can adapt individually and develop effective data processes.

In the system design context, a description of mental activities in information processing concepts is preferable because it is compatible with the concepts used for design of the control equipment and of the information processing and display. For this purpose, the human data processes can be described in terms of data, models, and strategies. *Data* are the mental representations of information describing the system state that can be represented on several levels of abstraction, which in turn specify the appropriate information coding for the data presentation. *Models* are the mental representation of the anatomic or functional structure of the system at the level of the search. The model can be directly related to the appropriate display formats. *Strategies* are here taken as the higher-level structures of the mental processes, and they identify the set of models, data, and tactical process rules that together may serve a particular task.

Within this framework, the operators' actual performance can be described in terms of *tactical rules* used to control the detailed processes within a formal strategy, together with the rules representing the shifts between the formal strategies that take place. During real-life performance, such shifts frequently occur when difficulties are met in the current strategy, or when information is observed that indicates immediate results from another strategy. Because the shifts are controlled by detailed person- and situation-dependent aspects, they result in the very individual course and impulsive appearance of the mental processes that were discussed in the previous chapter.

Typology of Diagnostic Search Strategies

In general, the diagnostic task implied in supervisory systems control is a search to identify a change from normal or planned system operation in terms that can refer the controller to the appropriate control actions. Such a diagnostic search can be performed in two basically different ways.

A set of observations representing the abnormal state of the system–a set of symptoms–can be used as a search template to find a matching set in a library of symptoms related to different abnormal system conditions. This kind of search will be called symptomatic search. On the other hand, the search can be performed in the actual, maloperating system with reference to a template representing normal or planned operation. The change will then be found as a mismatch and identified by its location in the system. Consequently, this kind of search strategy was called topographic search in the previous chapter.

The difference between the two kinds of search strategies is related to a basic difference in the use of the observed information. Every observation implies identification of an information source and reading of the content of the message. By symptomatic search, reference to the identity of system state is obtained from the message read; by topographic search, reference is taken from the topographic location of the source, whereas the messages are subject only to good/bad judgments that are used for tactical control of the search.

Symptomatic Search

The topographic search is performed by a good/bad mapping of the system through which the extent of the potentially "bad" field is gradually narrowed down until the location of the change is determined with sufficient resolution to allow selection of an appropriate action. The domain or level in the abstraction hierarchy in which the search is performed will vary. The search can be

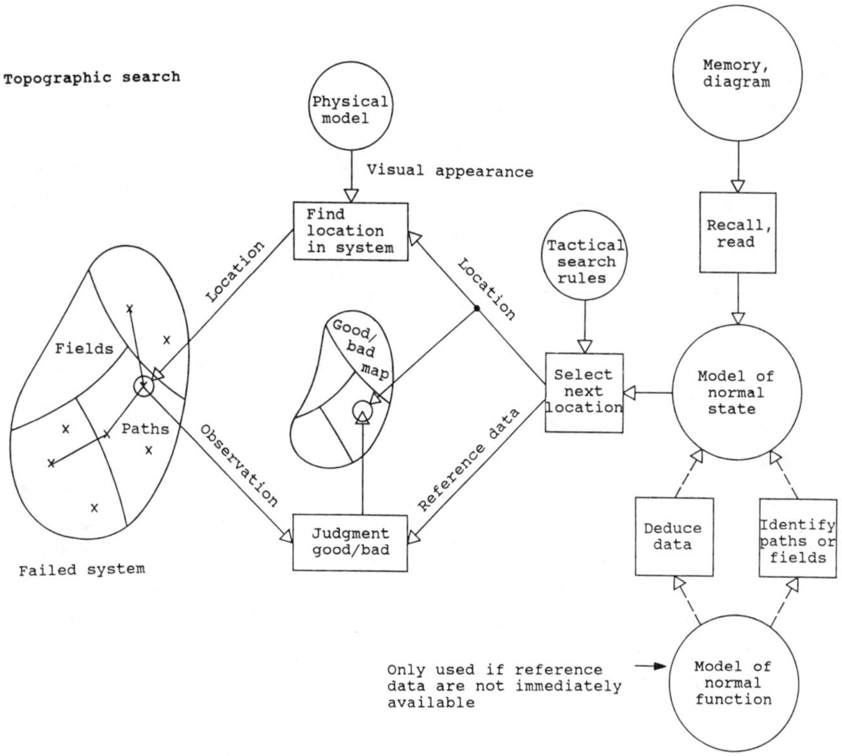

Figure 6.1. Information flow map illustrating the topographic search strategy, which is based on good/bad judgments of variables along a path or of patterns related to a field. [Reproduced from Rasmussen (1981) with permission from Plenum Publishing Corp.]

performed directly in the physical domain but, in most cases, the search is a mental operation at a level of abstraction that depends upon the immediate goal and intention of the controller and upon the form of the reference map or model available. Also, the resolution needed for the final location depends upon the actual circumstances.

The topographic strategy is illustrated by the information flow graph of Figure 6.1. The main elements of the strategy, which will be considered in more detail, are the model of the system used to structure the search; the kind of data used to represent the actual, failed plant state and the normal, reference state; and finally, the tactical process rules used to control the search sequence.

The topographic search is performed as a good/bad mapping of the system, which results in a stepwise limitation of the field of attention within which further search is to be considered. The search depends on a map of the system that gives information on the location of potential sources of information for which reference information is available for judgments. The map is a model that identifies the potential sources of observations relative to the topology of the physical system itself.

The search sequence is based on a set of typically heuristic rules serving to limit the necessary field of attention. If different external functions can be related to separate internal parts or subsystems, a good/bad scan of external functions effectively identifies the internal field for further search. If a faulty input/output relation is found, the related causal route should be searched, e.g., by the half-split heuristic. In the pure form, the tactical search decisions are exclusively based on the one bit of information obtained from the good/bad judgment of the individual observations.

The information available in observations is used rather uneconomically by topographic strategies because they depend only upon good/bad judgments. Furthermore, they do not take into account previously experienced faults and disturbances. Therefore, switching to other strategies may be necessary to reach an acceptable resolution of the search or to acquire good tactical guidance during the search. However, the topographic search is advantageous because of its dependence upon a model of normal plant operation, which can be derived during design or obtained by data collection during normal operation. Consistency and correctness of the strategy can therefore be verified and, because the model does not depend on models of malfunction, it will be less disturbed by multiple or "unknown" disturbances than strategies based on symptoms.

Symptomatic Search

Symptomatic search strategies are based on the information content of observations to obtain identification of system state, instead of the location of the information source. The search decisions are derived from the internal relationship in data sets and not from the topological structure of system properties. In principle, a search is made through a library of abnormal data sets–"symptoms"–to find the set that matches the actual observed pattern of system behavior. The reference patterns can be collected empirically from system maloperation or derived by analysis or simulation of the response of the system to postulated disturbances. Furthermore, reference patterns can be generated on-line if the diagnostician has a functional model available that can be modified to match a current hypothesis about the disturbance.

When the diagnosis is performed in the data domain by a search through a library of symptom patterns, it has no logical relation to system function. The result is directly the label of the matching symptom pattern, which may be in terms of cause, effect, location, or the appropriate control action itself. Depending upon the structure of the controller and its memory, the search can be a parallel, data-driven *pattern recognition*, or a sequential *decision-table search*, as illustrated in Figure 6.2.

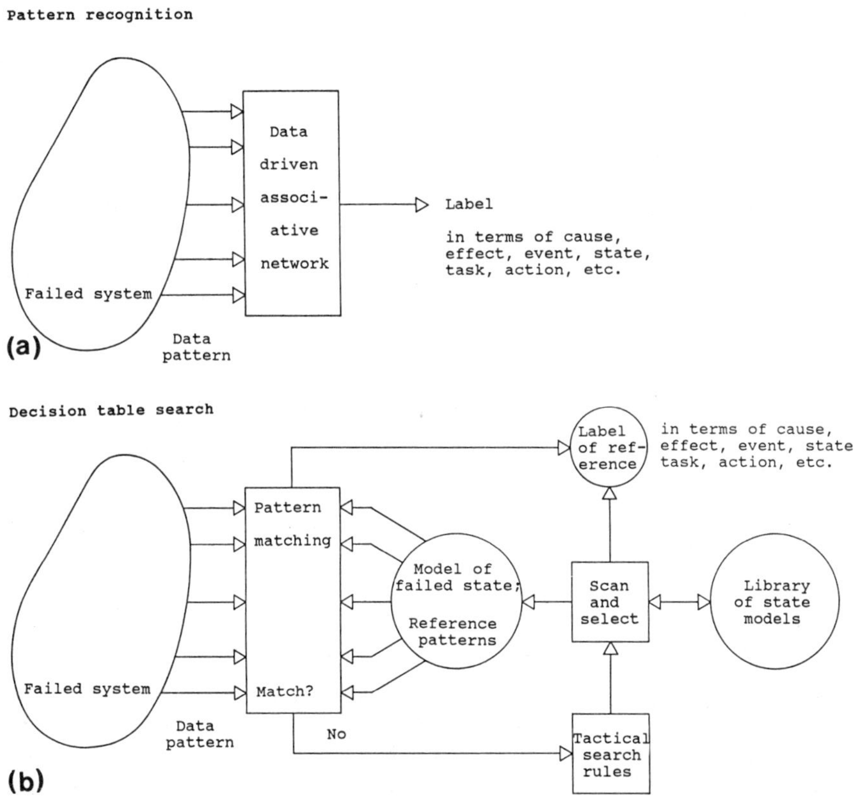

Figure 6.2. Information flow maps for symptomatic diagnosis based on (a) pattern recognition or (b) decision-table search through a library of symptoms. [Reproduced from Rasmussen (1981) with permission from Plenum Publishing Corp.]

Pattern recognition plays an important role in human diagnosis; it can effectively identify familiar system states and disturbances directly, but it is also frequently used during topographic search to guide tactical decisions. Recognitions are then typically based on more fuzzy or general reference symptoms in terms of generic fault patterns referring to types of function or physical part, such as noise characteristics, instability, or forms of nonlinearity.

Decision-table search depends upon a set of tactical rules to guide the search that can be based on probability of occurrence, a hierarchical structuring of the attributes (like Linné's generic system for botanical identification as used in field guides), or functional relations, stored as fault trees, etc. Human diagnosticians would probably use a kind of mental decision table search for verification of more ambiguous recognitions.

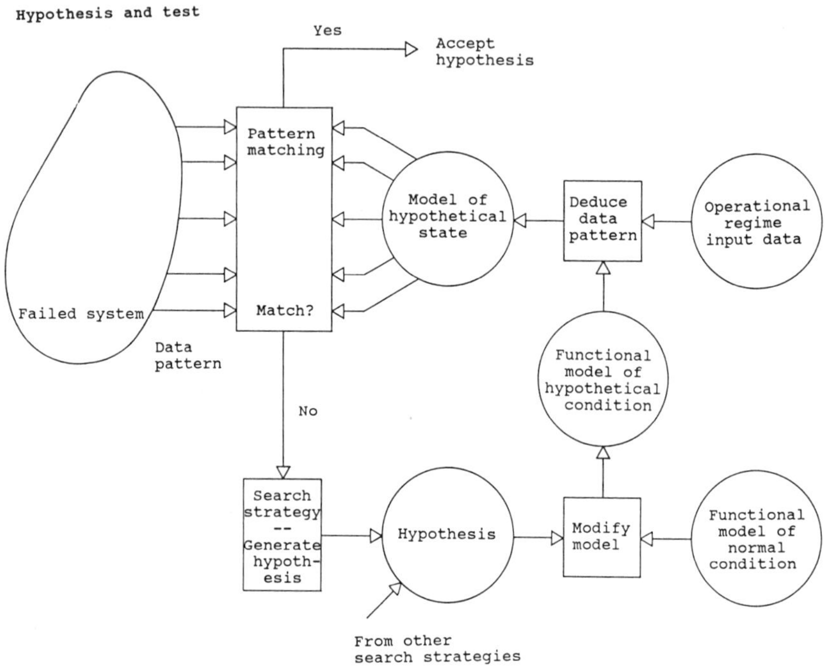

Figure 6.3. Information flow map for symptomatic search by hypothesis and test. The figure illustrates a conceptual test. In practice, the test may be performed by correcting the system according to the hypothesis and test for normal functioning, instead of modifying the model as shown. [Reproduced from Rasmussen (1981) with permission from Plenum Publishing Corp.]

If a search is based on reference patterns generated on-line by modification of a functional model in correspondence with a hypothetical disturbance, the strategy can be called *search by hypothesis and test*, as illustrated in Figure 6.3. The efficiency of this type of search depends upon the tactics of generating hypotheses. Typically, in human diagnosis, hypotheses result from uncertain topographic search or fuzzy recognitions.

Symptomatic search is advantageous from the point of view of information economy, and a precise identification can frequently be obtained in a one-shot decision. One serious limitation will be that a reference pattern of the actual abnormal state of operation must be available, and multiple faults and disturbances may be difficult to take into account. Reference sets must be prepared by analysis or recorded from prior occurrences, or the reference sets can be generated on-line by means of a functional model of the system that can be modified on occasion to simulate the abnormal system state in accordance with the current hypothesis.

Resource Requirements of the Diagnostic Strategies

The information flow maps of the diagnostic strategies shown in Figures 6.1-6.3 represent the structure of strategies only, together with information about necessary data-processing models. Because they do not contain information on the particular data-processing sequence that will be used in a specific situation, they can be used to represent a whole family of diagnostic processes typical of classes of diagnostic situations. This means that they are well suited to serve as the basis for interface design and human-computer task allocation. Task allocation implies the matching of the resource requirements from the different strategies with the resources available to the system designer, the human operator, and a process control computer for coping with the diagnostic task during different situations. In this section, the resource requirements of the different strategies will be discussed in some detail in order to illustrate the kind of design decision that must be made.

Consider first the topographic search, which is basically a good/bad mapping with reference to a topographic map of the system supplying reference data for judgment. The proper level of topological representation and hence the nature of reference data and observations depend on the immediate control task. For overall control of coolant inventory and energy flow in a power plant, a representation of the mass and energy flow topology will serve the purpose of state identification in terms of deviations from specified operational state. For fault location in a system, a topological representation at the level of physical function identifying components and their individual function is necessary. With respect to resource requirements, an important feature is that the search is based on a model of the normal functional topology and state, which is independent of the actual situation, although, as we have seen, different representations of the topology may be appropriate for different situations. However, a proper set of reference models can be prepared in advance by the systems designer, either analytically or by simulation, or it can be developed by measurement upon the operating plant itself. Using the process computer for data collection and analysis can lead to a currently updated reference model of normal state, not only serving as a reference during acute diagnostic situations, but also for detecting gradual changes in performance.

The key process of the diagnostic search is the judgment of discrepancy between the actual state and the target, i.e., the normal state, at the relevant level of functional representation. The judgment itself is a simple process. However, preparing information for the search implies information processes posing a high demand on processing and memory capacity: the proper reference model must be selected and updated from information about the operational regime and the actual system configuration with respect to switching and valving. Then, measured data from the system should be integrated with information that can be used for the good/bad judgments of the relevant levels of topological representations. This means that the physical variables measured in the system are only directly useful at the lowest levels of physical function. For search implied in supervisory control of process plants at, for instance, the level of mass and energy topology, all the available data from the system should be used to derive information about flows and levels in the mass/energy flow structures and balances. In a similar way, equipment that makes it possible to trace information flow, rather than signal flow, through an information processing system, is useful for troubleshooting purposes. The reference for judgment can, for instance, be recorded "signatures," i.e., normal information patterns, along the paths. Since this preparation of information for topographic search implies well-structured processes with considerable requirements for processing and memory capacity, the task is well suited for computers.

The topographic search can be based on various search tactics that will lead to different resource requirements. The search can be based on judgment of the magnitude of state variables directly. If judgments along a flow path of information, energy, etc., are used, the search is similar to the topographic search used for electronic troubleshooting discussed earlier. If judgments of sets of variables related to specific functions or parts in the system are used, the search is similar to the functional search discussed for troubleshooting. Given a proper topological map and proper reference data, this search can be very efficient with human judges. In complex process systems, proper topological displays with data arranged for immediate visual comparison can be generated by computers, leading to low resource requirements on the part of the human operator. However, search by direct judgment of the magnitude of variables can lead to difficulties, for instance, in case of feedback loop effects, and it can be difficult to distinguish between disturbed and disturbing functions. Therefore, more elaborate search tactics should be considered, in particular when computer support is possible.

Very effective strategies can be developed if they are based on relationships among variables that are invariant with the detailed operational state and, therefore, with the magnitude of the individual variable. A simple case of search based on state-invariant relationships has been mentioned in Chapter 5; a technician used an ohmmeter to find a faulty resistor. Search by relationships is more complex if a topographic search in a flow topology is based on conservation laws by checking mass or energy balances rather than flow variables. Lind (1981) has proposed a systematic search strategy based on inferences in the flow topology and has used programs inspired by artificial intelligence (AI), such as PROLOG, for computer-based diagnosis. Because his approach aims at identification of changes at a high level of abstraction in terms of deviations from normal flow topology, dynamic properties of the system have not been considered. A systematic topographic search at the level of physical function and based on a reference in the form of the normal dynamic properties of equipment or functions has been proposed by Sheridan. This approach is, however, only aimed at identification of the location of the initial disturbance (Sheridan, 1981; Sheridan et al., 1982).

In the case of diagnosis of a system that is not in operation, a very effective topographic search can be performed by forcing the system through a sequence of properly chosen operational states, *test states*, which affect various functions of the system in different ways and in properly selected combinations, and for which reference models can be prepared. Administration and evaluation of the test results then depend on logical, combinatorial arguments. Computer-based support for such diagnosis in complex systems like the Apollo space system has been used (Furth et al., 1967; Wohl, 1981).

All the topographic strategies depend on search with reference to a *model of normal function* and are therefore well suited for identification of disturbances that are not empirically known or that the designer has not foreseen. However, some of the more elaborate search procedures are based on evaluation of quantitative relationships in sets of variables and/or logical combinatorial arguments for which human operators have only limited resources. Computer-assisted topographic diagnosis therefore seems to be a promising area for further development.

The symptomatic search strategies are, in general, very information-economic, but they can only be applied when a library of symptom procedures is at one's disposal. Depending upon the source of reference patterns available, the symptomatic search takes different forms with very different resource requirements.

State identification and diagnosis by data-driven pattern recognition represents, in particular, the human operator's perceptive recognition of familiar patterns, and will be the normal way to associate a situation to the related human activities. The requirement for conscious data processing is low, provided the necessary data are available in a format suited for immediate perceptual recognition. In automatic equipment, data-driven recognition is used in the traditional, hard-wired safety system, which releases protective actions when a predetermined set of measured variables exceeds the trip limits.

Decision-table search depends on the availability of a library of symptoms that are used as templates in a search for a match with the observed data. In the table, the symptoms are entries to information related to the identified state. This information can be in terms of the root cause of a disturbance, statements of the task to pursue, or directly in terms of the procedure to follow. The decision table can be present as a kind of association matrix in the memory of a human operator or as a proper decision table in a process computer.

Whether the output from the table takes the form of a state identification, a task prescription, or a procedure directly depends on experience from the operator's previous cases or the control designer's ability to predict the state and make decisions on the proper response. That is, the decision table will be a record of the results of previous decisions by the operator himself or by the designer. How much of the decision sequence of Figure 2.1 the designer can prepare in advance depends on the invariance across occurrences of the class of situations considered, and on the designer's ability to foresee the relevant conditions for decisions. What is not included in the designer's decisions, and

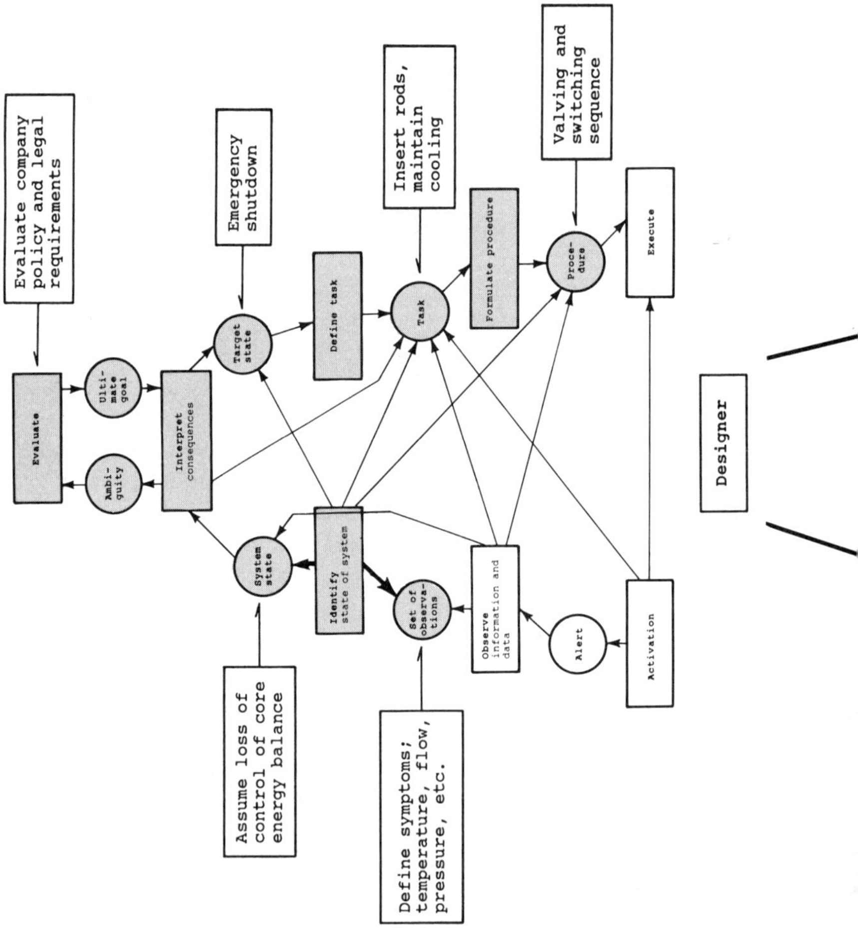
Evaluate company policy and legal requirements
Emergency shutdown
Insert rods, maintain cooling
Valving and switching sequence
Evaluate
Ulti-mate goal
Ambi-guity
Interpret consequences
Target state
Define task
Task
Formulate procedure
Proce-dure
Execute
System state
Identify state of system
Set of observa-tions
Observe information and data
Alert
Activation
Designer
Assume loss of control of core energy balance
Define symptoms; temperature, flow, pressure, etc.

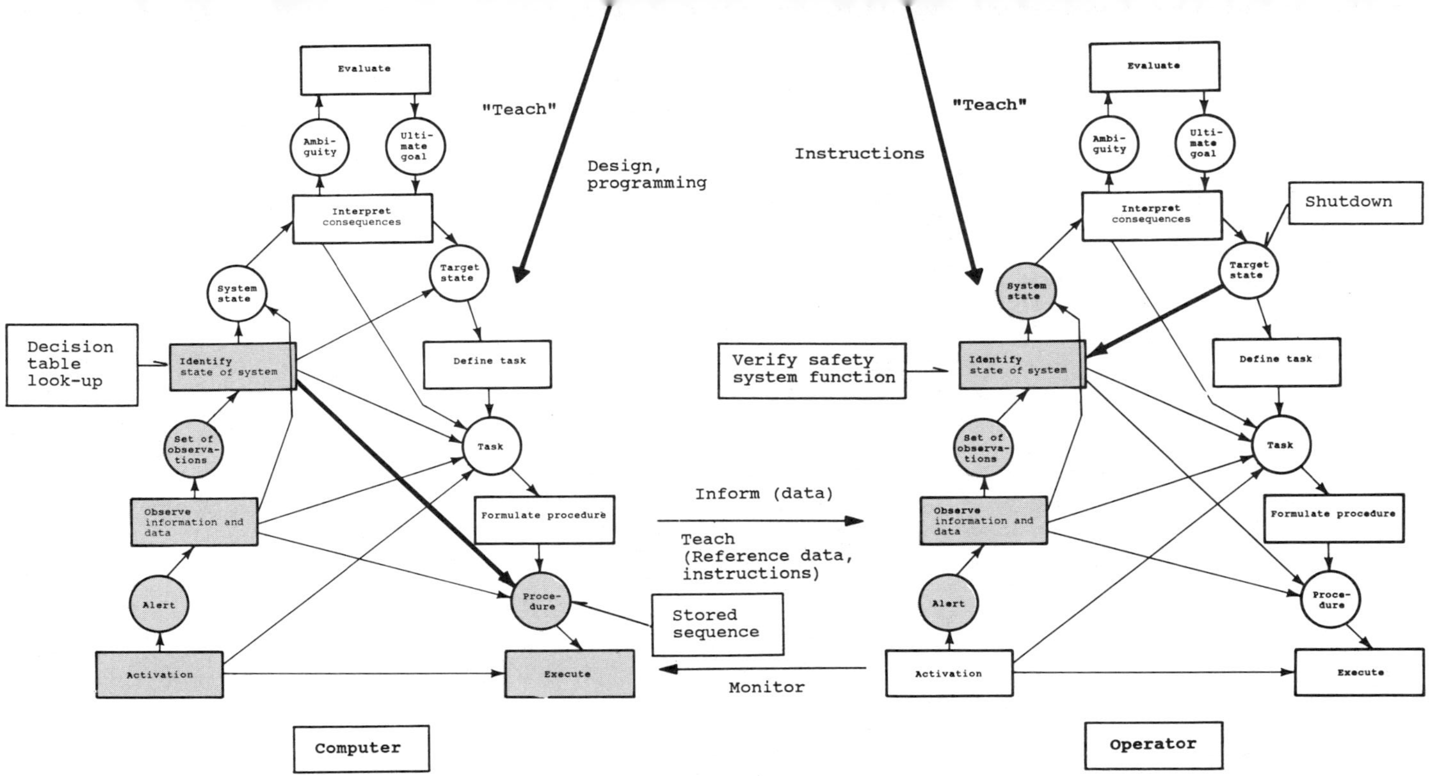

Figure 6.4. The designer, the operator, and the process computer are all parts of a control decision. The figure illustrates their roles (shaded) in an automatic safety shutdown of a nuclear reactor, in which they act "in parallel," each with a diagnostic task based on different strategies. The figure only serves to illustrate the cooperative decision making; for details of the decision ladder, see Figure 2.1.

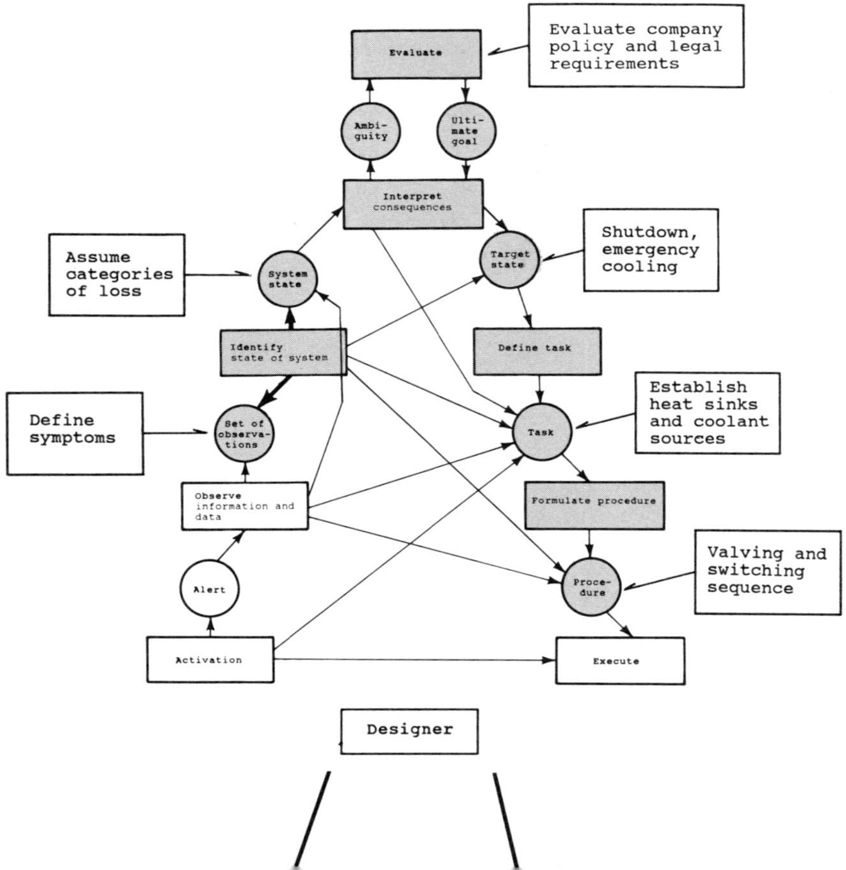
Evaluate
Evaluate company policy and legal requirements
Ambi-guity
Ulti-mate goal
Interpret consequences
Shutdown, emergency cooling
Target state
Assume categories of loss
System state
Identify state of system
Define task
Establish heat sinks and coolant sources
Task
Define symptoms
Set of observa-tions
Observe information and data
Formulate procedure
Valving and switching sequence
Proce-dure
Alert
Activation
Execute
Designer

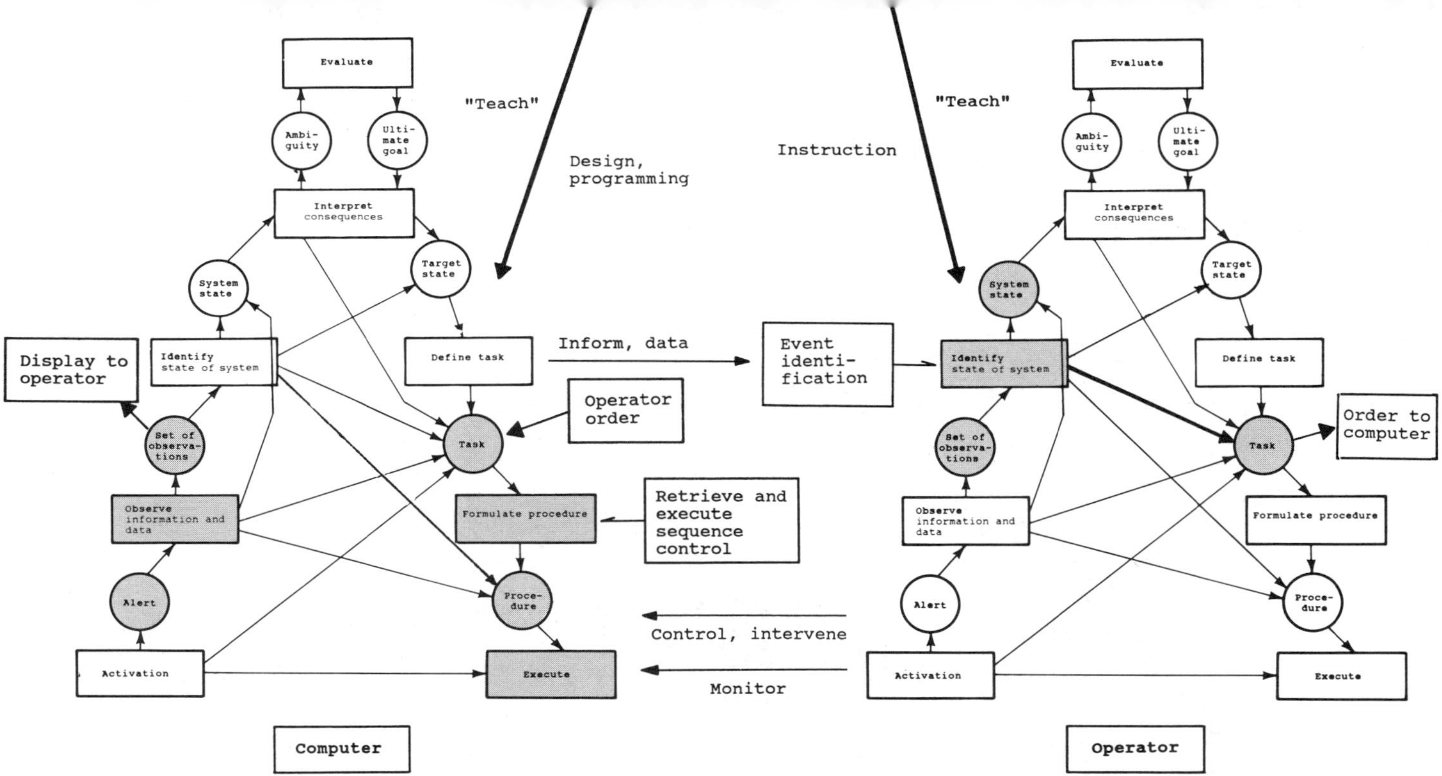

Figure 6.5. This figure illustrates the roles of the designer, the operator, and a computer in a control decision related to protection of a nuclear reactor during a loss-of-coolant accident that has not been fully automated; i.e., the operator and computer act in series. The designer has automated a repertoire of protective sequences, but left the diagnosis to an operator. In addition to the decision function, the designer, operator, and computer support each other in different inform-teach-learn functions, as discussed by Sheridan (1982).

hence in the decision table, must be left to the operator's decision during the actual situation. This can lead to rather a complex cooperation between designer, computer, and operator, such as shown in Figures 6.4 and 6.5. Furthermore, if the designer does not have a well-structured approach, the role of the operator and a computer may vary considerably with the plant situation, and the question of responsibility will be very difficult to settle.

The requirements for processing and memory capacity for decision-table storage and use is very high because of the need to represent a high number of special cases. Therefore, only for situations in which operators meet regularly during work or at training simulators is it possible to rely on human memory and know-how. Less frequent but well-structured situations can only be covered if external support is available for information storage and retrieval. Traditionally, operators in process plants have been supported by event-related, written procedures that leave the event identification to the operators themselves. However, in some industries like nuclear power, a trend is found toward support in the form of symptom-based procedures including the diagnostic phase. Likewise, several attempts have been made to automate state identification by means of computers that store the designer's analysis of fault patterns in decision-table format (Lees and Andow, 1981). However, it is difficult to evaluate the response of such systems to complex situations. For major disturbances of system operation, it is generally a problem for a designer to generate proper reference symptoms, because this requires analysis or simulation of the system outside the normal operating ranges and configurations.

A functional model for analysis or simulation of *abnormal* function is also a major requirement of the *search by hypothesis and test*. A number of problems are related to the development of a functional model for this purpose. First,the model must represent the abnormal function of the system outside the operating ranges and the configurations that are normally considered by the models used for system design and control studies. Such models are structured as a network of quantitative, mathematical relations among physical process variables, and these relations must be known for all the relevant states of maloperation. For on-line generation of symptom patterns during a disturbance, the model can be implemented by a process computer and updated according to the actual hypothesis. An advantage would be that the model could not only be useful for hypothesis testing, but also to predict, in accelerated time, responses to intended control actions. However, in addition to the problem of model range, it can be difficult to update the model to correspond to the prevailing hypothesis, in that there will generally be a very complex mapping from physical changes of the system onto the corresponding changes in model structure and parameters. In particular, this is the case when model updating reflects a hypothesis formulated by a human operator, because this hypothesis will not be in terms compatible with a quantitative computer model.

Functional reasoning by humans will not be based on a model in terms of relations among variables, but in complementary terms of objects and components, as well as states and events, as discussed in Chapter 10. Mapping of a hypothesis using this kind of model is much simpler, but the symptom patterns supplied by human reasoning may be too uncertain and qualitative for a reliable test against measured data. Ultimately, human operators may therefore choose to test their hypothesis by correcting the systems accordingly. While this may be an effective strategy in a workshop environment, it may lead to further complication of the situation in process plant control if the hypothesis is wrong.

For human-computer cooperation in a task such as search by hypothesis and test, the complementary nature of functional models suited for use by the two parts should be considered, and the techniques used in AI for computer implementation of qualitative reasoning (Brown and DeKleer, 1981; Williams et al., 1982) may be useful for design of intelligent interfaces.

Training Requirements

In general, operator training is taken care of separately from systems design by training specialists. This is most unfortunate, particularly in considering future systems where interface and display formats are matched to specific tasks and situations. To do this, the designer has to consider the mental strategies that will be adopted by operators and the resources required in terms of background knowledge. Such background knowledge must include the relevant models of system properties, which very much depend on the strategy chosen for a given task. Training can be considered a way to guide operators toward development of the mental strategies and models chosen for interface design. Design of training schemes must therefore be integrated with the other aspects of system design.

There has recently been a trend in this direction. Shepherd et al. (1977) have studied the influence of the different content of training programs upon diagnostic performance. They found that operators trained by the rules obtained from experienced operators were superior in diagnosis of faults not previously encountered compared with operators trained in plant theory. These in turn were superior to operators trained by practicing diagnosis from symptom patterns. These differences can be readily explained, inasmuch as the different training methods support the use of different strategies (i.e., topographic search, hypothesis and test, and recognition, respectively), which have different characteristics with respect to identification of new fault types.

Topographic search in abstract flow structures is very similar to rule-based search in "context-free" networks described by Rouse et al. (1980). It is interesting to note the observation of Rouse that the rule-based model describes the context-free strategies reasonably well but breaks down in context-dependent experiments. This is probably because the context initiates shifts to symptomatic strategies that are depending upon the individual subject's prior experience. Rouse and coworkers have later (e.g., Rouse, 1983) defined expert diagnosticians as those who know when it is time to leave the symptomatic search in favor of topographic search in unfamiliar cases. An approach to training of general diagnostic skills, based on training in varying problem contexts, which supports the use of topology-related rather than symptom-related strategies, seems to be promising (Rouse, 1982a,b).

The conclusion here is that the design of training schemes for operators must be based on identification of the mental models and strategies that will be effective for the tasks the operators are supposed to perform. These models and strategies must then be used for integrated design of communication interface and training schemes. An illustrative review of the resource requirements of the various diagnostic strategies is given in Table 6.1.

Table 6.1. Difference in Resource Requirements of the Various Diagnostic Strategies Which Makes Possible a Resource-Demand Matching by Proper Choice of Strategy. Features not filled in are either not typical, being dependent on circumstances, or not relevant. [Reproduced from Rasmussen (1981) with permission from Plenum Publishing Corp.]

	Strategy			
Performance factor	**Topographic search**	**Recognition**	**Decision table**	**Hypothesis and test**
Time spent	–	Low	–	–
Number of observations	High	Low	–	Low
Dependency on pattern perception	–	High	–	–
Load upon short-term memory	Low	Low	High	High
Complexity of cognitive processes	Low	Low	–	High
Complexity of functional model	Low	–	–	High
General applicability of tactical rules	High	–	–	Low
Dependency on malfunction experience	Low	High	–	Low
Dependency on malfunction preanalysis	–	–	High	–

Chapter 7

Design of Supervisory Control Systems

The introduction of new technology such as computer-based information processing necessitates a reconsideration of the basis for design of industrial control systems. A gradual updating of previous design in response to operational experience is very likely to lead to systems that are suboptimal and unnecessarily complicated. As an example, the use of computers to analyze alarm signals in process plant control rooms has been introduced to avoid overload from the high number of alarms, which in turn are a result of the traditional one-sensor-one-indicator technology with normal range monitoring of measured variables individually.

A supervisory control system is basically a feedback system with the task to monitor the actual operating state of the system and to keep it within the specified target domain. As is the case for any feedback control system, the design must be made without knowledge of the detailed nature of the disturbances with which the system should cope. The design must be based on the identification of the categories of disturbances and control tasks, and on consideration of the functional resources necessary for the system to be able to cope with these categories. This is particularly the case for systems with highly adaptive features, such as supervisory control systems that include human operators.

This section considers how the categories and models discussed in the previous sections can be used to formulate the control requirements of the systems and identify the possible strategies that can be used as control algorithms by the supervisory control system. This kind of analysis is necessary to bridge the gap between the traditional tools and methods of control system designers and the recent results of research within cognitive science. Unless the control systems

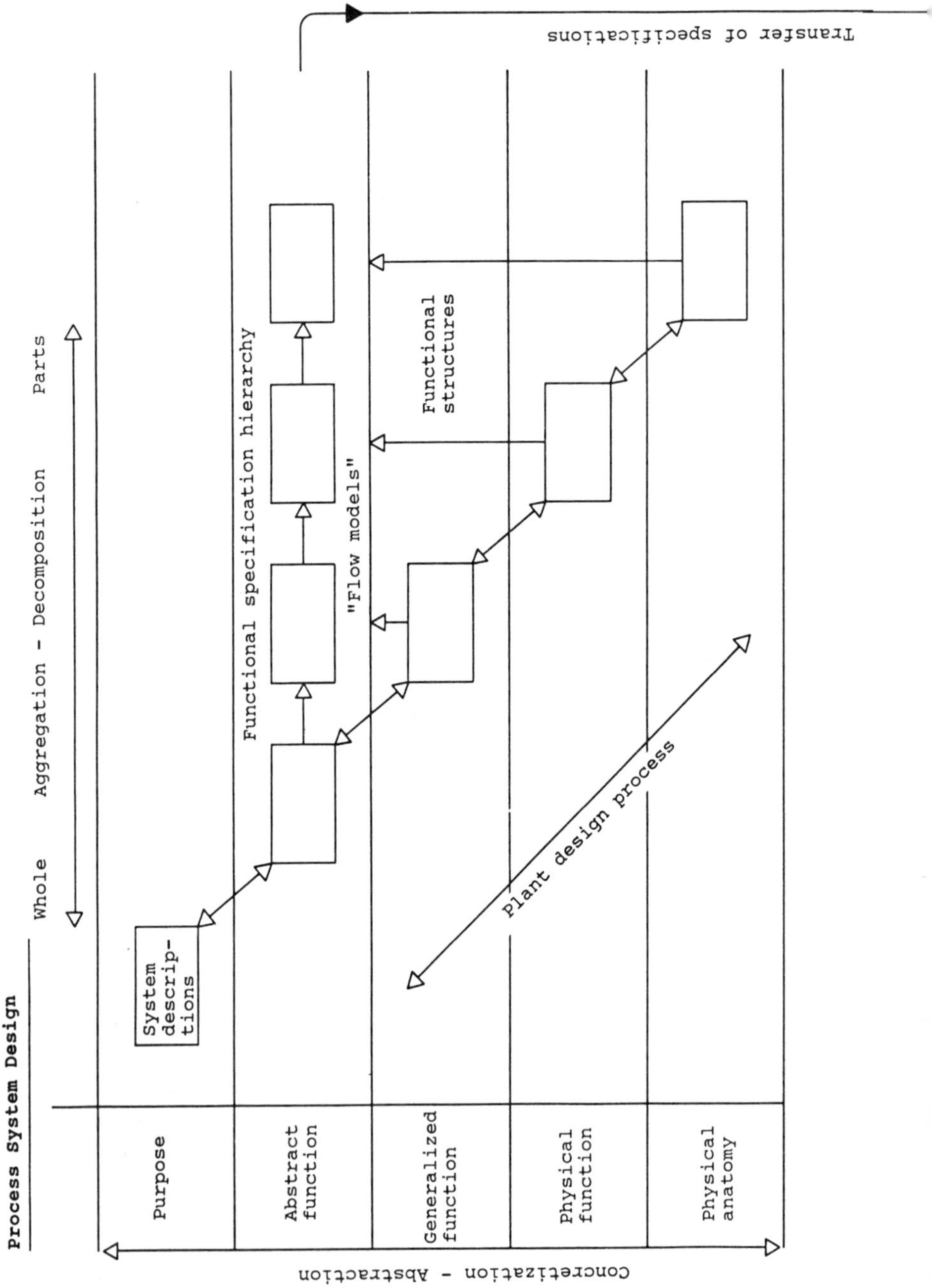
Process System Design
Whole
Aggregation - Decomposition
Parts
System descriptions
Functional specification hierarchy
"Flow models"
Functional structures
Plant design process
Purpose
Abstract function
Generalized function
Physical function
Physical anatomy
Concretization - Abstraction
Transfer of specifications

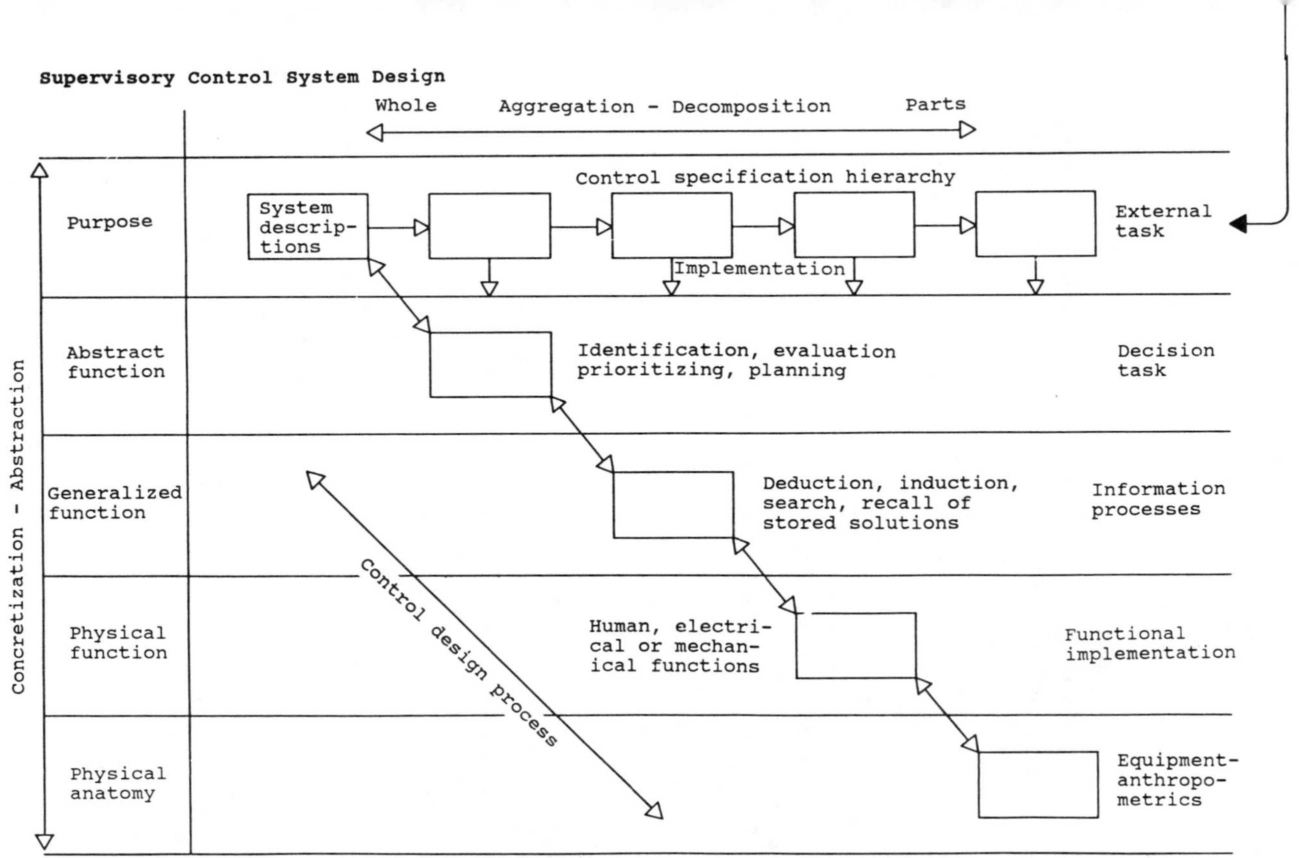

Figure 7.1. The design of a process plant involves an iteration simultaneously in a whole-part and an abstraction-concrete dimension. As concrete details emerge, the related possible functions must be constrained by a control system. Systematic design of a supervisory control requires that the functional specifications and the control requirements are identified in terms of device-independent information processes, before implementation and man-computer task allocation are considered.

designer also identifies the set of information-processing strategies that will meet the control requirements, he will not be able to ask meaningful questions of cognitive science concerning human capabilities and preference for the interface design. A possible way to structure such a systematic design process aiming at introduction of advanced information technology is discussed in the following pages.

During the design of the physical process system, the functions of the system and their physical implementation are developed by iteratively considering the properties of the system at the various levels of the abstraction hierarchy and in increasing degree of detail, as shown in Figure 7.1. As the degree of physical detail increases during design, so does the number of degrees of freedom in the functional states. The constraints necessary to maintain operation within the specified target domain must therefore be identified by the designer. In this way, the desired states of functions and equipment will be identified during design at different levels of abstraction, and the necessary control constraints will be identified in terms of the conceptual framework related to the individual levels. In general, a skilled designer will immediately be able to identify suitable and familiar control system concepts. However, when new technology is to be introduced for control implementation, it is necessary to develop a consistent description of the total control system function in a uniform, device-independent terminology. This means that the control requirements identified at the different levels must be considered together in terms of device-independent information processes.

The control requirements of the system are derived from the necessary relations between the possible system states, the specified target states, and the actions required to maintain control of the system. For a device-independent formulation of the information processes involved in supervisory control, the decision sequence of Figure 2.1 can be used. Depending upon the control task allocation, the different phases of the decision sequence will be performed by the designer himself, the system operator, or the process computer.

For well-structured situations, for instance, start-up and shutdown sequences, the designer can carry through all the decision steps, except the execution itself, and store the design in a decision table in a computer or in a set of operational instructions for an operator. For many other situations, the designer may be able to plan proper tasks and procedures, for instance, for system protection, but not be able to foresee the related disturbed system states. In this case, diagnosis must be performed on-line by operator and computer in various degrees of cooperation. Finally, situations occur where all decision phases rely on the human operator, but because the necessary information must be supplied to the operator, his preferred strategies and related information needs must be considered when designing the interface.

In all three cases, there will be an intimate cooperation between designer, operator, and computer during supervisory decision making, as previously illustrated in Figures 6.4 and 6.5. Even when the computer performs automated

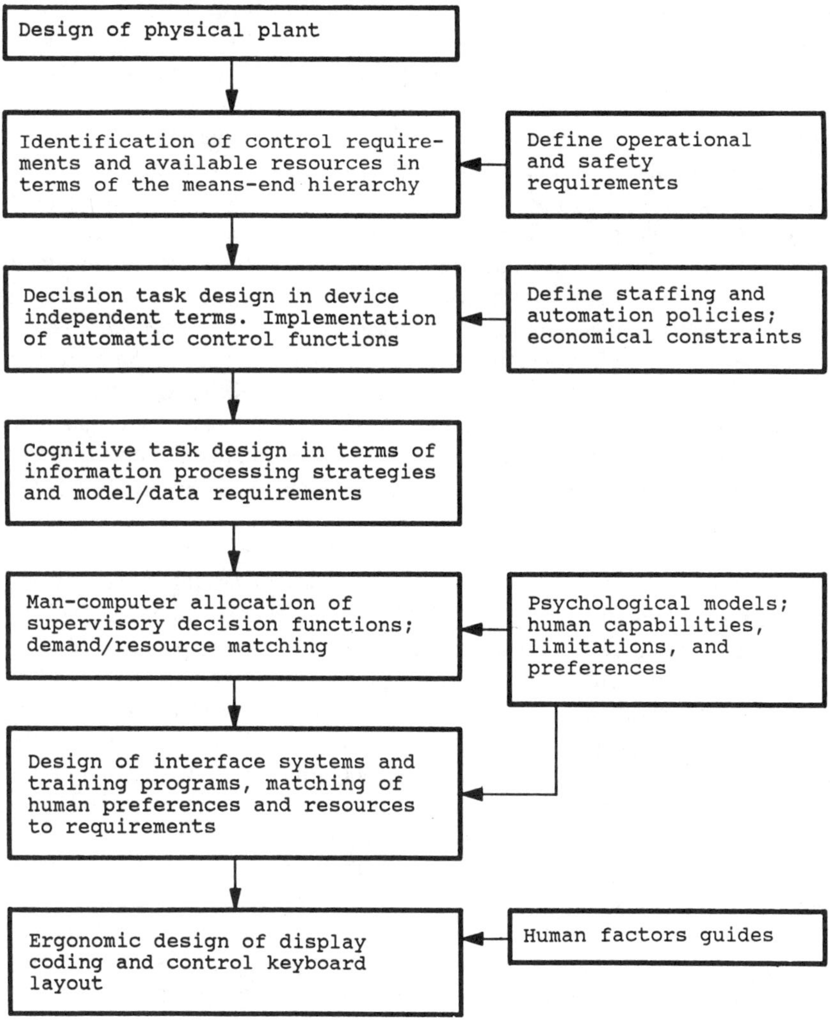

Figure 7.2. Phases in a systematic design of supervisory control systems.

control actions based on the designer's analysis, the operator will have to monitor performance. Therefore, he must be supplied with information about system states and about the designers' intentions in order to be able to verify the decisions taken by the computer, through use of his own preferred decision strategy.

In conclusion, design of supervisory control systems based on advanced information technology involves several distinct phases related to the concepts discussed in the previous chapters and shown in Figure 7.2. This figure illustrates a clear, well-ordered design sequence for sake of clarity. The real design task will, however, be a complex iteration back and forth between different phases. Nevertheless, this will not affect the main lines of the present argument.

First, the *control requirements* of the system are identified in terms of target states related to different operation regimes, together with the relevant categories of possible normal and disturbed states. This identification must be performed at each level of the abstraction hierarchy to develop the context of the control task, including identification of task specification (why) from the level above, and of the available resources (how) from the level below the particular task level.

Second, the *decision task* necessary to meet the control requirements is analyzed in terms of a device-independent framework like the decision ladder of Figure 2.1. During the analysis, the conditions for the different decision phases are evaluated in the light of the manning and automation policy of the particular operating organization and the safety requirements imposed. Then, one evaluates the extent to which stereotyped bypasses in the decision sequence can be analyzed and implemented as automatic functions in order to simplify the decision task during actual operations for well-structured and foreseen situations. This phase determines the functions to be taken care of by the automatic process control system.

The third stage is a *cognitive task analysis and design* serving to identify and describe the possible information-processing strategies that can be used for the various phases of the decision sequence involved in the supervisory control task.

Although the automatic control system typically takes care of all the functions during normal operation, routine start-up and shutdown sequence control, and protective actions in the more stereotyped and foreseen disturbances, the supervisory control is dependent on the operating staff, who monitor the automatic control functions, adjust for optimal coordination, and take over control when disturbances (or maintenance operations) can no longer be adequately managed by the automatic system. To an increasing degree, computer-based information systems will be used to support operators, and the supervisory control task will lead to complex cooperation between computers and humans, who will share the decision task itself, as well as functions for mutual support and monitoring. In addition to these supervisory functions, the cooperation between humans and computers may involve such functions as mutual teaching/learning (Sheridan, 1982), which should be considered in the cognitive task analysis.

For each phase of the analysis, evaluation, and planning stages of a supervisory control decision, several information strategies will be possible, both when involved in the control itself and when serving for monitoring and support. As has been discussed in previous sections, these strategies will have very different resource requirements that match computers and human operators differently. Therefore, the roles of operators and computers as controller and independent support and monitor are quite likely to vary during a decision task.

The fourth step in a systematic design will therefore be to evaluate the match between the resources available for implementation by means of human operators and computers, and to perform a *cognitive task allocation*. For this, a set of models of human information capacity and limitations is necessary, together with knowledge of the subjective task formulation and performance criteria that are likely to control the choice of strategy in the actual situation. The nature of these aspects has been illustrated for a troubleshooting task in Chapter 5. The need is clearly not for one comprehensive model of human operator performance, but for several different models related to different traditional fields of psychological research.

This is also the case for the fifth phase, the *design of an interface system* , that has the information coding and display formats that will serve to match human preferences in a way that activates the proper strategies. During this phase, the design of a training scheme matching the requirements of the cognitive tasks must also be considered.

An important aspect of this approach is that only after considering the first three phases will the system designer be able to pose questions to cognitive psychology and training research. The cognitive task analysis is not a psychological problem, but a question of identification of the data-processing strategies that can serve the control requirements of the system. These strategies must be formulated during control system design, but in system-independent terms that make it possible to relate capability requirements to the resources offered by humans and computers. Only then can system designers use the models of human data processing emerging from cognitive science.

Chapter 8

The Human as a System Component

A discussion of models of human behavior immediately raises the distinction between qualitative and quantitative models. Among engineers, qualitative models are frequently considered merely to be premature, descriptive models, which after further work will develop into or be replaced by proper quantitative models. However, this is not necessarily the case. In several respects, the two kinds of models have different and equally important roles for analysis and prediction of performance. This difference in significance is related to the distinction between categories of behavior and the members of such categories, i.e., the specific behavior in particular situations. Bateson (1979) discusses this distinction in detail with reference to Whitehead and Russell's logical types:

> ... there is a deep gulf between statements about an identified individual and statements about a class. Such statements are of *different logical types*, and prediction from one to the other is always unsure. [Bateson, 1979, p. 46.]

The fact that "the generic we can know, but the specific eludes us" (Bateson, 1979, p. 45) has different implications depending upon the purpose of the modeling effort. For systems design, qualitative models will serve important purposes if they are able to predict the category of behavior that will be activated by different possible interface configurations and display formats. The model will then support the choice of an interface design that will activate a category of behavior having limiting properties compatible with the functions allocated to the human operator. In a way, research on human performance in order to support system design should not focus on modeling actual performance in existing environments, but possible performance in optimal, future systems, as has been discussed by Sloman (1978) in a philosophic context. Qualitative models identifying categories of behavior and the limiting properties of the related human resources will serve designers well in the design of systems that allow humans to optimize their behavior within a proper category. Compare this with Norman's (1981) arguments for the importance of considering the proper mental image for design of "friendly" systems and the need for a profession he calls "cognitive engineering." The distinctions between models of categories and of particulars have different implications depending also on the cognitive level of behavior considered. At the level of sensorimotor behavior, we are considering highly trained people, similar to experimental psychologists' "well-trained subjects" who have adapted to the particular environment. In this domain, models of optimal human performance are mainly models of the *behavior of the environment* as seen through the human. Therefore, generic quantitative models of human performance in well-structured tasks can be–and have been–developed at this level of performance. At the level of improvisation and problem solving, we are dealing with individual reactions to unfamiliar situations, and models will be more a question of qualitative matching of categories of system requirements with human resources. For unfamiliar tasks, these resources depend on a specific person's subjective preferences, experience, and state of training. In this context, training means supplying people with a proper repertoire of possible behaviors for unexpected situations, and qualitative models matching categories will be highly effective. Until recently, training of system operators has not been based on models of human performance compatible with those used for systems design. However, the explicit use of qualitative models for matching categories of system requirements and human resources for planning of training programs by Rouse and his coworkers (Rouse, 1982b) has turned out successfully and proves the value of qualitative models.

To be useful, qualitative as well as quantitative models must reflect the structure underlying the mental processes, i.e., the internal or mental *models*, the kind of *data* dealt with by the processes, and the *rules* or *strategies* used to control the processes. In addition, the models must reflect the limits of human capabilities so that human "errors" also are modeled properly.

This question of also modeling errors properly leads directly to the issue of analog, parallel-processing models versus the sequential digital models of human information processes of the AI community. Can holistic human perception, for instance, be properly modeled by the sequential "production rule" systems? In the present context of models for system design and evaluation, the fundamental question appears not to be whether a model is implemented for experimental evaluation by means of one or another physical information processing system, but whether or not there exists a theoretical framework formulated independently of the tools for experimental implementation. This framework must have a one-to-one correspondence to human psychological mechanisms, their processing limitations and error characteristics. If such a separation between model and implementation was maintained, many of the arguments between psychologists and AI researchers (Dreyfus, 1972) could be circumvented. An implication of this point is that computer programs based, for instance, on the "production systems" of Newell and Simon (1972) cannot in general be accepted as theories unless they adequately represent limiting properties and error characteristics of the human processes. Proper representation of the failure properties of human information processes will be difficult, for instance, if holistic perception is modeled by sequential scene analysis. Therefore, proper evaluation of a model requires analysis of instances when the model breaks down, rather than a search for correspondence with human performance in successful instances. This is the essence of Simon's statement (Simon, 1969):

> A thinking human being is an adaptive system.... To the extent he is effectively adaptive, his behavior will reflect characteristics largely of the outer environment... and will reveal only a few limiting properties of his inner environment....

Successful performance does not validate a model; only tests of its limits and error properties can do this.

Levels of Abstraction in Models of Human-System Interaction

Models of humans can be developed at any level of the abstraction hierarchy discussed in Chapter 4 (see Figure 4.3). Models at the various levels are useful in different professional contexts, and an identification of the scope of the present discussion of human capabilities will be appropriate. For design of decision support and interface systems for supervisory control, the interest in modeling is focused on human resources for information processing. Capabilities and limitations are, however, influenced bottom-up in the hierarchy, and goals and intentions are determined top-down. It is therefore necessary to define and describe the interrelation between the information-processing models and the models at the adjoining levels.

The interaction between a system and a human should be considered at all levels, as illustrated in Figure 8.1. The figure shows how influence upon the human at any of the lower levels can modify capabilities and limitations at the abstract information-processing level. According to this view, an information processing model should be considered in relation to a social system model at the level above and a psychological model at the level below, as indicated in Figure 8.2. At the psychological level, we have the interaction between the system, viewed as the human work environment, and the psychological human properties, such as the dependency of capability upon interface ergonomics and information coding, upon motivational factors related to boredom and fatigue, and the dependencies of process and performance criteria upon workload, stressors, etc.

At the level of information processing, we may consider the interaction of the actual task and the human considered as an information processing component in

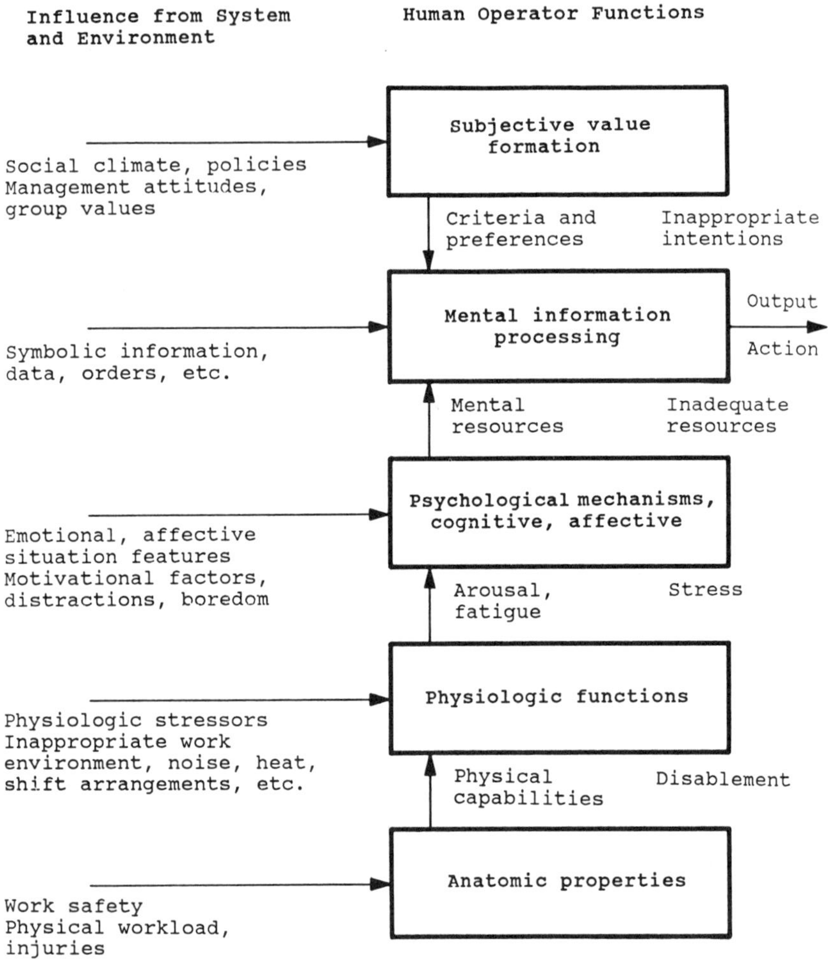

Figure 8.1. The diagram illustrates the complex interaction in a man-machine system that controls the mismatch features in an error situation. [Reproduced from Rasmussen (1982) with permission from Elsevier Science Publishing Co., Inc.]

the system. This is clearly a simplistic distinction, but it is necessary to have a framework in which engineering and psychology can formulate a consistent interface, the more so in that the research methodology at the two levels may have distinct differences. Models of human information processes are typically related to specific tasks and information processes and are based on functional models, whereas psychological models of human cognitive and affective properties and subjective preferences depend more on factor and scaling-type analysis and emotional state models. Interface systems based on modern information technology are necessarily designed from functional considerations

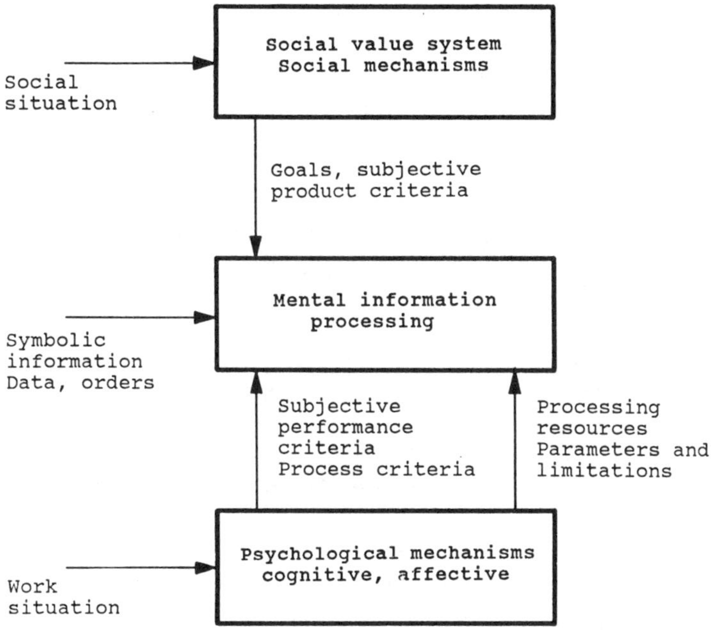

Figure 8.2. For man-machine system design, a model of humans in terms of information processes is useful. However, the definition of concepts interfacing such a model to social, affective, as well as cognitive, psychological mechanisms is important.

of the information processes they have to support. Whether the design chosen will in fact match human preferences and therefore activate the intended information processing mode depends on many more global and situational features and, today, the question is typically solved empirically at a very late stage in the life of the system. It will therefore be advantageous to formulate a consistent set of concepts and parameters that defines and describes their interaction.

To illustrate the kind of models we need, we will briefly reconsider some of the discussion of mental strategies from the previous sections. In those sections, the emphasis was on a framework for discussion of normative strategies that can be used for design consideration. In the following section, the emphasis will be on description of the strategies that humans actually use.

Description of Strategies in Real-Life Tasks

In order to study mental models and strategies, we have to refer to a specific task scenario. In our context, however, the problem is to derive guidelines for the design of human-machine interfaces suitable for a large repertoire of tasks related to a specific system. Therefore, reference can only be made to a generalized description of typical task sequences. This description must be based on analyses of the performance of skilled operators in their normal work situation. An effective way to distinguish their internal processes in this situation is to obtain and analyze verbal protocols supported by observations, interviews, etc. As discussed in Chapter 5, the efficiency of skilled performance results from the ability to compose the process needed for a specific task as a sequence of familiar subroutines that are useful in different contexts. This implies the existence of links in the sequence at standard key points, or "states of knowledge," which are characteristic of the specific skill. The data process stops at such links, the mode of processing, and frequently the level of abstraction change. To study and identify the processes, a description of the activity must be structured according to such key points. In familiar situations, the subroutines depend upon subconscious or nonverbal processes, and verbal protocols from such situations typically express a sequence of such states of knowledge without any indication of the connecting data processes. However, the structure of the sequence of verbal statements gives important information on the strategies used for control of mental processes. Only in unfamiliar situations, when operators have to improvise and generate new subroutines, will verbal protocols reflect data processes at a more detailed level. The verbal statements of the task sequence reflect "states of knowledge" in different categories representing plain observations, properties of the environment, goal, plans, etc., i.e., knowledge at different levels of abstraction and associated with different roles in task strategies. It follows that the type of data process needed to derive one state of knowledge from the preceding one depends upon the categories of these states.

In a situation where the operator uses rational data processes based on knowledge of the causal structure of the system being controlled, the different states of knowledge follow each other in an apparently logical order, which has been represented in Figure 8.3. In this figure, the steps of the basic sequence in a way form a "ladder of abstraction."

To be able to predict the behavior of a system and its response to operator actions, its internal state must be known to the operator. The process necessary to derive this knowledge from the observations is basically one of abstraction and of induction. In practice, this process can be a complex set of elementary processes including search, hypothesis generation and test, model transformation, performance evaluation, etc.; see the discussion of diagnostic tasks in Chapter 3. The top of the ladder dealing with the determination of the target state for the system from the present state and of the overall operational goal implies a prediction of system behavior, performance evaluation, and value judgments. In general, the operator will not climb to this level of abstraction. The goal will be implicitly defined by the situation; the internal model will be controlled by the subconscious value/goal system and a specific system state will immediately associate to a target state or a task. When a target for the system state is chosen, the operator must determine the task, i.e., the proper control actions to perform,

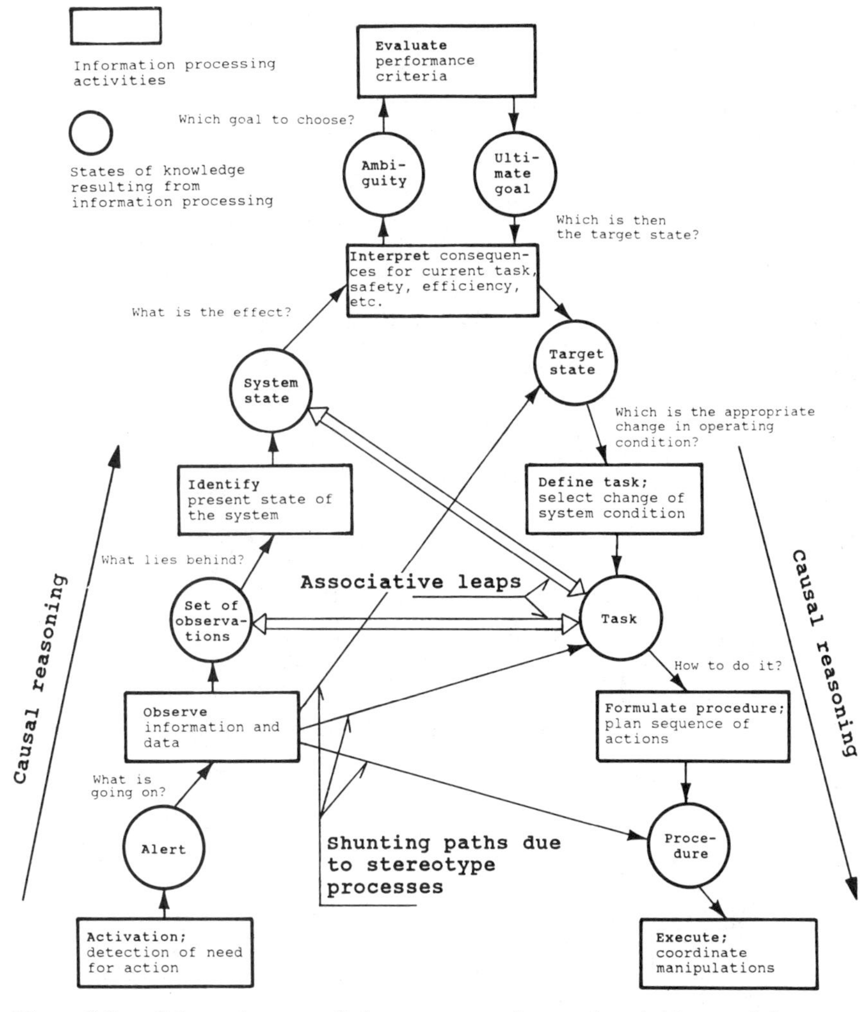

Figure 8.3. Schematic map of the sequence of mental activities used between initiation of response and the manual action. Rational, causal reasoning connects the "states of knowledge" in the basic sequence. Stereotyped mental processes can bypass intermediate states. Associative leaps can connect any states of knowledge. [Reproduced from Rasmussen (1980) with permission from Academic Press, Inc.]

by a process of deduction. The operator now moves down from the higher levels of abstraction to more specific technical details, identifying the possible changes of the plant, such as switching, valve manipulation, or adjustments, that will lead to the desired plant state.

In the basic sequence of mental processes based on knowledge of the causal

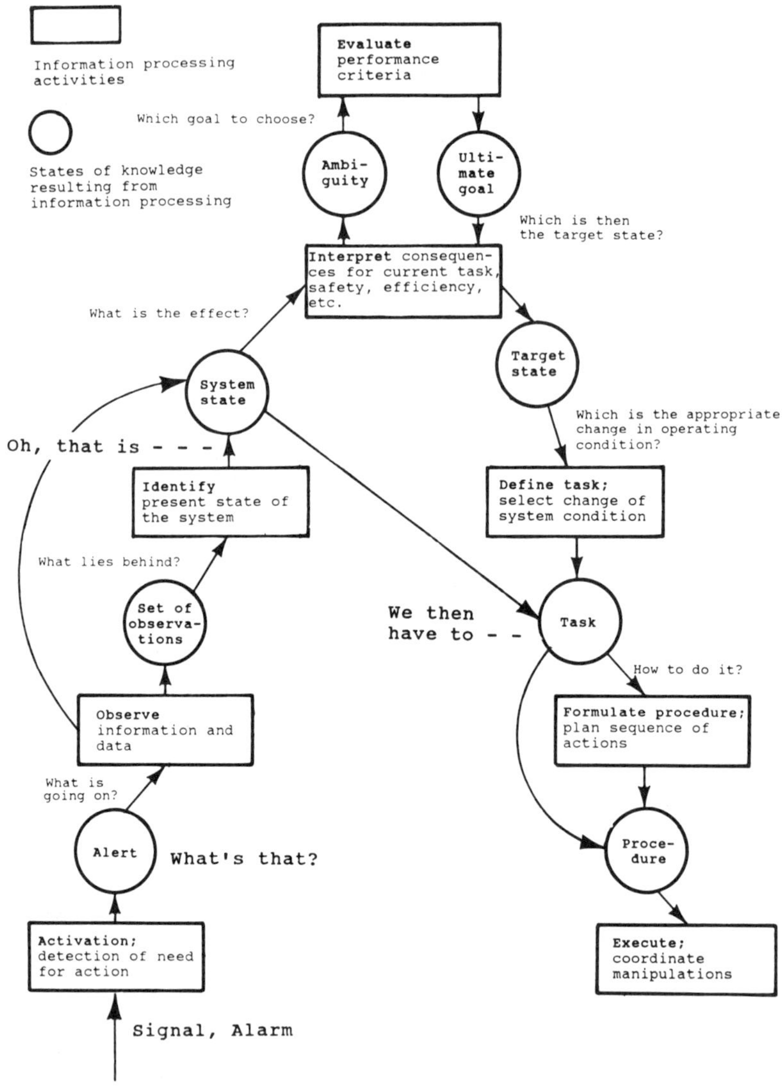

structure of the system, the different states of knowledge may be expected to follow each other in a logical sequence. This is not necessarily the case if the data processes are stereotyped or proceduralized–if they follow "cookbook" prescriptions, which may be the case in familiar or preplanned activities. Then there may be no logical connection between subsequent states of knowledge, and some of the higher-level processes of the basic sequence can be bypassed as shown in Figure 8.3. Regardless of which of these two cases apply, the operator must follow a sequential, conscious data process when a problem is recognized;

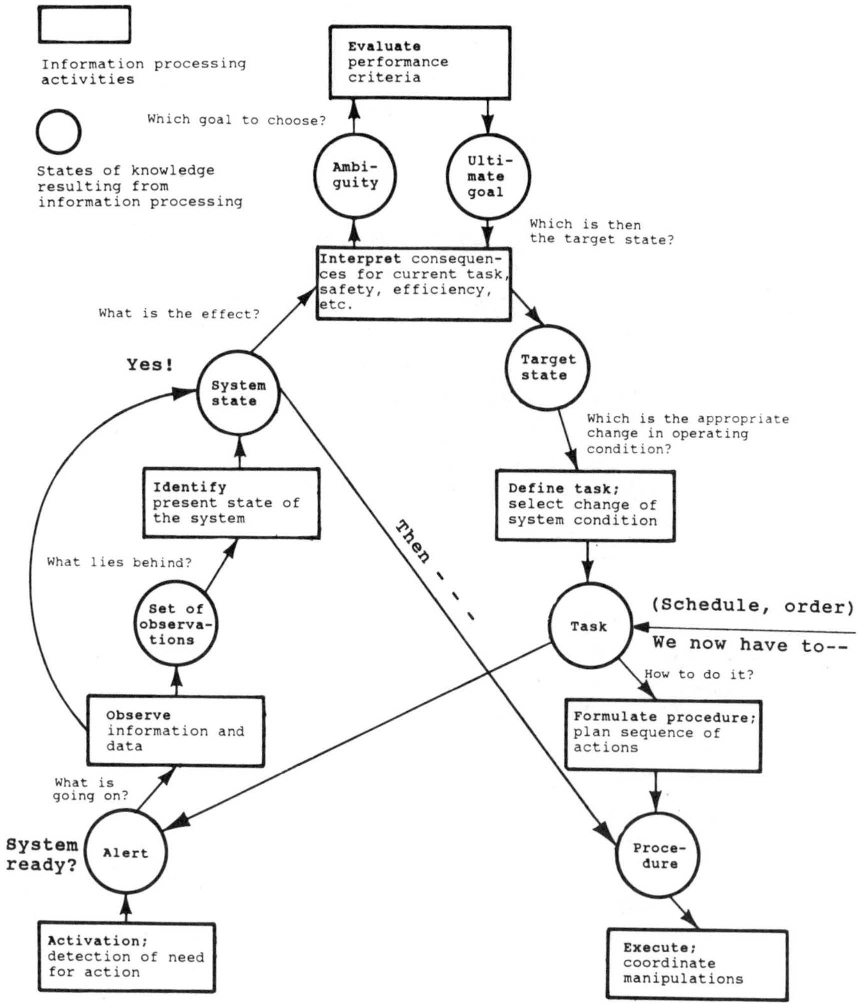

Figure 8.4. Two typical samples of routine sequences from skilled operator performance. [Reproduced from Rasmussen (1980) with permission from Academic Press, Inc.]

---

i.e., an uncertainty needs to be resolved. However, immediate associations may lead directly from one state of knowledge to the next. Such direct association between states of knowledge is the typical process in familiar situations and leads to very efficient bypassing of low-capacity, higher-level processes. Such associations do not follow logical rules; they are based on prior experience and can connect all categories of "states of knowledge." Some typical examples of the associative links found in verbal reports from skilled operations are given in Figure 8.4.

In considering the diagnostic task in Chapter 3, it was mentioned that the circularity in the relationship between the determination of the level of diagnosis to consider and the goal to pursue leads to the need for an iterative approach to the decision process. This will be the case in any real-life problem with potentially multiple causes and goals; consequently, the decision ladder is not a model of the decision process itself, but rather a map useful to represent the structure of such a model.

In supervisory control of a physical process, system dynamics will constrain this iteration and the time spent. In less structured situations–for instance, in managerial decision making–the iteration back and forth between data collection, evaluation, and selection of criteria for the ultimate decision may be spread over considerable time and involve many persons; see, for instance, the case studies of Cyert and March (1963). The involvement of several persons with different perceptions of goals may be a support for the necessary iteration. Research on individual subjects' information seeking behavior in multiple-cause situations indicates a pronounced tendency to focus on a single, linear causal relationship, rather than to explore a more complex causal net (Shacklee and Fischhoff, 1982). This is in good agreement with the tendency to use focusing strategies under the influence of the law of least resistance for electronic diagnosis. Neither in our nor in Shacklee and Fischhoff's experiments, however, were the subjects exposed to the potential drive from multiple goals toward more thorough analysis that generally is effective in managerial decision making.

In everyday tasks, evolution has fit the human for an efficient interaction with a time-space environment where high-capacity parallel processes take care of the lower functions of the ladder of abstraction. An abundance of redundant information supports feature extraction and the formation and synchronization of the internal world model. In process control, however, an operator controls a physical process from which only preselected information is presented and typically as symbolic representations of individual variables. Designers often suppose that the operator uses rational, conscious processes–even in lower-level tasks–that would lead to complex time-sharing between tasks at different process levels. When studying process-operator performance, an important task is to identify the repertoire of ingenious tricks developed by operators to avoid this situation. For example, an operator's interpretation of information will resemble the focusing strategies discussed by Bruner et al. (1956) and will be based upon a model of normal state, leading to separate judgments based on individual observations (see Chapter 5). The operator relies on intuitive judgments and expectations which, owing to the nature of the subconscious world model, may be based on representativeness rather than on a rational foundation (Tversky and Kahneman, 1974).

Human data processes in real-life tasks are extremely situation- and person-dependent, and a detailed process description is difficult to obtain. This is even the case when an unfamiliar situation forces the operator into a process of detailed functional reasoning from which a good verbal protocol can be obtained.

In most cases, several different strategies are available by which the operator may solve the problem. These strategies may all be rational, but differ according to different performance criteria. However, in actual work situations, these different strategies are typically not followed systematically. As discussed in Chapter 5, the mental processes seem to follow the law of least resistance, and a shift between strategies will take place whenever a momentary difficulty is encountered in the current strategy or special information is recognized that makes another strategy look more promising. See Figure 8.5, which illustrates how models of formal, systematic strategies can be used to describe the strategies actually used.

Even when it is possible to obtain detailed verbal protocols from a single-task sequence, it may be very difficult to identify general features of the data processes because of such spontaneous shifts between different strategies, shifts that are initiated by minute details in the situation. One cannot see the wood for the trees. A simultaneous analysis of several process descriptions (e.g., from verbal protocols) in terms of categories of data, models, and procedural rules supports a more efficient identification of formal strategies; see Chapter 5. Another detailed analysis of the individual case can then lead to identification of the leaps between formal strategies. A description of what the operator will do in a specific situation can then only be obtained in terms of a set of typical, systematic strategies and the heuristic rules controlling the transitions between these strategies. Such a model could be quantified in terms of a Markov process if adequate data could be collected.

The discussion of models of humans as system components leads to the following conclusion for the requirements of a model: the model should reflect human behavior in various domains, including habitual sensorimotor acts, know-how, and tricks of the trade, as well as rational decision making. It should not necessarily predict the detailed mental processes in a given task, only the categories of possible strategies and the human match to their resource requirements, together with the performance criteria that humans adopt for choice of strategy.

Outline of a Model of a Human Information Processor

The main emphasis of the present discussion, then, is to evaluate and interrelate the available models of human information processing that are compatible with the needs of designers of cognitive systems. It is, however, also necessary to consider models of human behavior at the psychological level of cognitive and affective mechanisms in order to develop models at two levels that are compatible and therefore can be the basis of demand-resource match considerations.Some general commonsense considerations will demonstrate that the model we need for design of real-life systems has to incorporate human abilities that are normally studied within quite separate research paradigms.

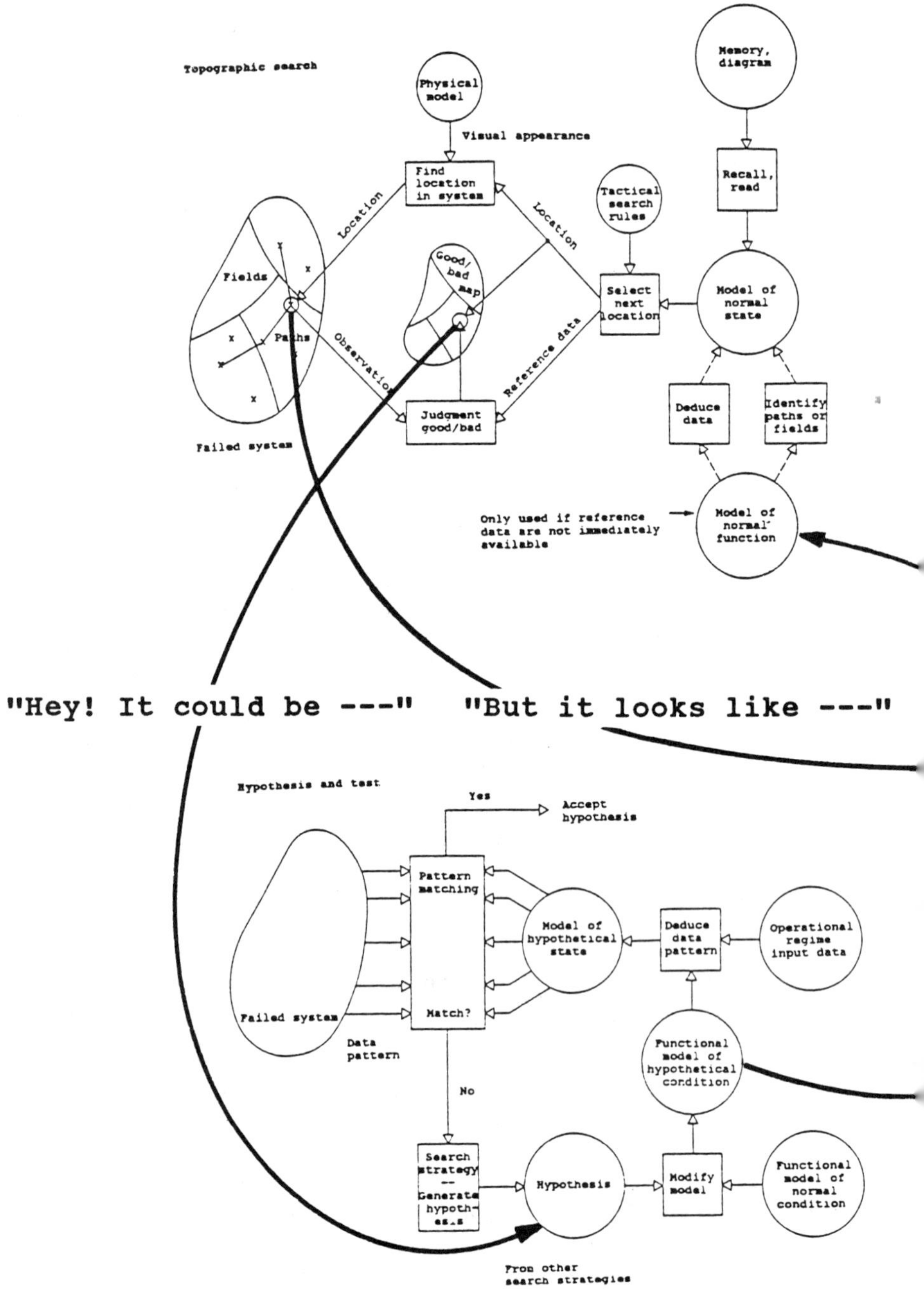
Topographic search
Physical model
Visual appearance
Memory, diagram
Find location in system
Recall, read
Tactical search rules
Location
Location
Fields
Good/bad map
Select next location
Model of normal state
Paths
Observation
Reference data
Judgment good/bad
Deduce data
Identify paths or fields
Failed system
Only used if reference data are not immediately available
Model of normal function
"Hey! It could be ---"
"But it looks like ---"
Hypothesis and test
Yes
Accept hypothesis
Pattern matching
Model of hypothetical state
Deduce data pattern
Operational regime input data
Failed system
Match?
Data pattern
Functional model of hypothetical condition
No
Search strategy -- Generate hypotheses
Hypothesis
Modify model
Functional model of normal condition
From other search strategies

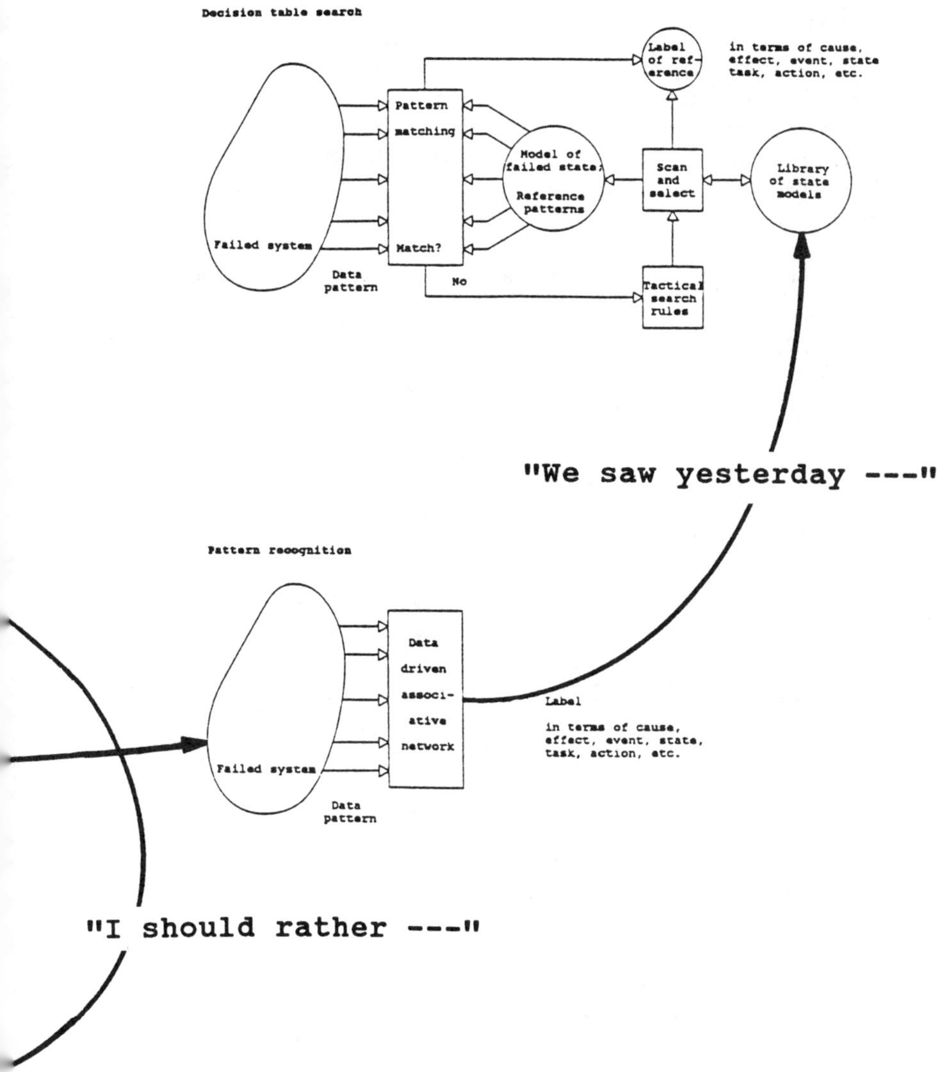

Figure 8.5. Examples of information flow maps of formal, systematic strategies found in a diagnostic task. The actual sequence consists of pieces of such strategies because of the frequent "jumps" between them. For details of the diagnostic strategies, see Figures 6.1, 6.2, and 6.3. [Reproduced from Rasmussen (1980) with permission from Academic Press, Inc.]

A Man Is Quite Simple–but in a Complex Way

In the discussion of the "sciences of the artificial," which is very relevant in the present context, Simon (1966) explores the hypothesis:

> A man, viewed as a behaving system, is quite simple. The apparent complexity of his behavior over time is largely a reflection of the complexity of the environment in which he finds himself.... his behavior will reflect characteristics largely of the outer environment (in the light of his goals) and will reveal only a few limiting properties of his inner environment, of the psychological machinery that enables him to think.

From this point of view, humans may be quite simple in some clearcut laboratory tasks, but the simplicity rapidly evaporates, when considering the models required for real-systems design. First of all, we are looking for models that represent precisely the situations when adaptability breaks down and the psychological machinery reveals itself. Furthermore, clear evidence exists that humans have several basically different internal modes of information processing, which are put into operation by different types of outer environment and tasks. These different modes draw upon mechanisms with different resource characteristics, such as processing capacity, speed, etc. Some simple considerations demonstrate the span of the problem. The actual output of a human interacting with a system is physical acts upon the system, and the model must reflect the fact that sensorimotor performance enables a human to drive a car through a town during rush hours at the same time as the conscious mind is occupied by reconsidering the problems of the office hours. It must also consider that humans may have difficulty remembering a telephone number from the directory long enough to dial it properly; at the same time they can recognize friends when met unexpectedly in a crowd, by features they are not able to describe verbally. A crucial feature to include in a model is that humans are not only responding to "input." They actively seek relevant information; they constantly orient their senses toward important features in the information available, and they respond to information present as well as information missing–they have expectations. The attitude behind the following discussion is close to the view of Gibson (1966); humans are not processing the information input from the environment–they are actively picking up the information that is relevant in the context of their current needs and goals. This context is formed by the general state of affairs in the environment and the individual person's role in it. Effective control of this information pickup is only possible if the individual has available an internal dynamic representation of the state of affairs, and an active, internal dynamic world model is therefore central in a model of human information processing.

Because an important coupling of the information processes to the environment is through physical actions, modeling should include this internal dynamic world model and its interaction with both perception and motor control.

It is generally accepted that high-capacity, parallel-processing neural networks are the mechanisms behind these functions that clearly take place below the level of conscious awareness. They can be consciously controlled, but typically this is done by conscious intentions that are not stated in terms of movements, but as higher-level goals–shut the door, drink your tea, etc.

From a systems point of view, we are interested in a discussion of the various human information-processing mechanisms and, in particular, their limiting properties related to the information processes used. In this respect, the human may be considered as a hierarchical system. The information processes necessary to coordinate the body with a dynamic environment are taken care of by the subconscious sensorimotor system, which can be viewed as a high-capacity, parallel-processing, pulse-density-coded analog computer. The activities of this system are subject to a higher-level control by a conscious, symbolic processor that has features similar to a programmed, sequential computer. If a computer metaphor should be used, humans are not like a digital computer, but like a complex computing center, interfaced to the dynamic world through a multivariable, continuous control system. On the following pages, the psychological mechanisms behind this metaphor will be discussed in more detail. The intent is not to give a comprehensive review of the current psychological research, but to discuss some key features in order to have a basis for discussing the various modeling approaches, and the extent to which they also adequately model the limits of human capabilities and the error characteristics that are important in the human-machine system context.

The model will be a very crude sketch. A human is far more than a mechanistic data processor, but in human-machine systems the designer has to consider people as versatile but predictable data processors, and the designer has to plan the data processing allocated to a human and to the instrumentation system as one integrated system. The human-machine interface should therefore be based upon compatible models of the data processing in humans and in instruments, and a model of humans drawing heavily upon engineering analogies may be fruitful in this respect. The general outline of the model is illustrated in Figure 8.6 and comprises two basically different subsystems. One is a programmable, sequential processor that operates consciously to the human. It has a rather limited capacity and speed, and its main task is to take care of the data handling in unique conditions calling for improvisations, rational deductions, and symbolic reasoning. It functions as a high-level coordinator or controller of the second part, the main data processor.

The main processor has many features of a distributed, parallel-processing analog computer with high processing capacity. It functions mainly subconsciously to the human, although the conscious processor may to some extent monitor the effects of the data processing. The functions are distributed in a highly interconnected network that is able to decode information from the sensing receptors, and to extract higher-level features or patterns. These control the external actions through several levels of coordinating functions in the motor

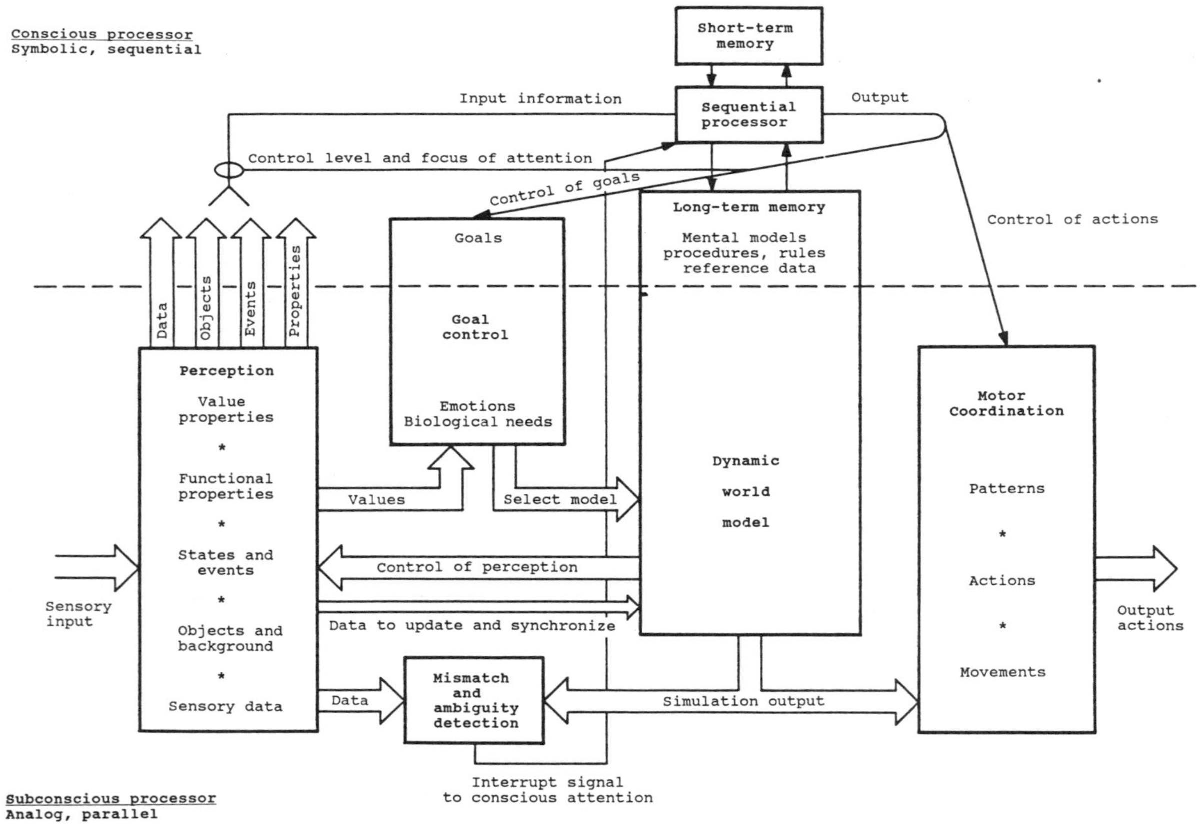

Conscious processor
Symbolic, sequential
Short-term memory
Sequential processor
Input information
Output
Control level and focus of attention
Control of goals
Control of actions
Goals
Goal control
Emotions Biological needs
Long-term memory
Mental models procedures, rules reference data
Dynamic world model
Data
Objects
Events
Properties
Perception
Value properties
Functional properties
States and events
Objects and background
Sensory data
Values
Select model
Control of perception
Data to update and synchronize
Sensory input
Motor Coordination
Patterns
Actions
Movements
Output actions
Data
Mismatch and ambiguity detection
Simulation output
Interrupt signal to conscious attention
Subconscious processor
Analog, parallel

Figure 8.6

system, which is an integral part of the processor. An important implicit function of this distributed network is a continuously dynamic modeling of the environment and the body. This internal model is updated or trained slowly, but it operates in real time and serves to control the actions and thus to synchronize the body with the dynamic environment.

Figure 8.6 is in no way intended as a representation of the organization of the underlying neuropsychological functions, but only as a map of the relationship among functions at the psychological level of Figure 8.1. The neurologic structure in terms of the abstraction hierarchy is represented at the next lower physiologic level.

Subconscious, Sensorimotor Functions

It is generally accepted that the basic nature of information generalization and the derivation of control information for the motor system, and hence the coding of the subconscious dynamic world model, are closely related to the basic needs of human ancestors during evolution. Survival has most probably been granted to those individuals who were best–by direct pattern recognitions–at quickly identifying cues in the environment related to immediate goals, such as approach, escape, eat, and don't eat, and who had the most efficient control of fast critical sequences during hunting, fighting, etc.

This dynamic interaction with the environment calls for a very efficient feature extraction and classification and dynamic coordination of the motor system with the environment. The receptors and the muscles of the body serve as input and output systems of a complex multivariable control system having a high data-processing capacity. The subconscious main processor may therefore be assumed to have evolved into a data-processing system with particularly high capacity in functions needed in the control of the body, i.e., when operating in a spatial, temporal domain. Correspondingly, Heidbreder (1949) has put forward the hypothesis that perception of concrete objects is the dominant mode of cognitive reaction related to the phylogenetic priority of locomotive and manipulative capabilities.

This function may be organized similarly in humans and in higher-order mammals, and it has been proposed that evolution has resulted in a hierarchical organization also of the neurologic structures in humans that are in essence

---

Figure 8.6. Schematic map of the human data-processing functions that illustrates the important role of the subconscious dynamic world model as part of a complex loop of interactions in conjunction with the perception and goal systems. The world model also forms the basis for a high-capacity, efficient feed-forward control of physical actions and serves as a reference for the mismatch detector that activates the conscious processor. [Reproduced from Rasmussen (1980) with permission from Academic Press, Inc.]

peculiar owing to the addition of a symbolic processor on top of three other neural levels we have in common with animals (Altman, 1978), i.e., the reptile, the vertebrate, and the mammal brain.

The Front-End Function: Perception

The front-end function is the extraction of meaningful features from the wealth of information available in the environment, which requires a processing system of high capacity.

In the task of controlling the body in the dynamic environment, the most important generalizations to be made from the input information will be to isolate and identify spatial entities and their temporal characteristics. This means to identify objects that are moving or are able to move and for which generic dynamic models are present in the internal dynamic modeling system. Furthermore, the generalizing function has to extract those generic features from the background–the scenery in which the dynamic actions take place–that are determining the behavior of the movable objects.

Research on visual perception has identified fixed programmed networks in several animals that generalize the visual information into this kind of dynamically determining feature. In the retina of a frog's eye, at least four types of preprogrammed detectors have been found: edge, bug, dimming, and event detectors (Lettvin et al., 1959). Similar detectors identifying form and movements of objects against the background are also found in higher-order vertebrates, although the feature extraction does not in all cases take place in the retina, but in preprogrammed parts of the nerve tissue at a higher functional level (visual cortex) (Hubel, 1963). A review of this research is given by Michael (1969).

Although programming of the human nerve tissue may be more flexible, Lashley (1942) expected to find similiarities. In his discussion of the cerebral organization of vision he notes:

> ... there is some reason for believing that generalization is one of the primitive, basic functions of organized nervous tissue... In spite of the enormous differences in structural arrangement of the visual systems of different animal classes, the functional activity seems essentially the same.... in the bird, the rodent and the man.... [From Lashley, 1969 edition, p. 236.]

This input-processing point of view has been continued in more recent cognitive modeling in terms of sequential scene analysis; see, for example, Minsky (1975). In these approaches, the information in the environment is analyzed in order to identify known objects that are then allocated functional properties and related to a propositional database. However, human perception seems to be a more active and direct process. Michotte's (1963) experiments on perception indicate a human ability to identify higher-level functional and causal relationships in dynamic light patterns that have no concrete object context. Patterns of spots are perceived in terms of A strikes, withdraws from, or pursues B, etc. Gibson (1966) has strongly advocated the view that perception is not based on processing of sensory input information, but on direct sampling of invariant features in information from the environment.

> It should now be clear that the brain does not have to integrate successive visual sensations in immediate memory. There is no necessary reason to suppose that the fixations have to be retained. The invariance of perception with varying samples of overlapping stimulation may be accounted for by invariant information and by an attunement of the whole retino-neuro-muscular system to invariant information. The development of this attunement, or the education of attention, depends on past experience, but not on the storage of past experience. [Gibson, 1966, p. 262.]

In reality, humans are probably able to use both "sequential scene analysis" and "direct perception" and both should be considered in the modeling effort, in that they will have quite different limiting properties and error characteristics. Gibson's views support the present arguments for the importance of the "internal dynamic world model," which can be viewed as the functional manifestation of the "retino-neuro-muscular attunement."

The relationship at the functional level between the world model and perception is sketched in Figure 8.6. An important feature to note is the loop formed by the world model and perception. The expectations of the organism represented by the world model will orient the sensory system visually, auditorily, as well as tactilely, toward the location of that source of information and that level of generalization where changes are expected or uncertainty in modeling exists. The perceptual system, on the other hand, supplies the information that serves to synchronize and update the model as well as the value features that, in some kind of interaction with the immediate needs and goals of the organism, activate the proper model.

The significance of the internal world model and its control of perception is that the whole system is sensitive to, "attuned" to, those invariant features that are essential to the actual goal in the context present. In fact, the environment is a continuum of information sources, and to cope with this great variety, the human information processor must be able to access this information in manageable higher-level structures. This means that information pickup is an active process that may be organized top-down in an abstraction hierarchy like that of Figure 4.1. The information available can be sampled for invariant features at a high abstraction level without having to perform a "bottom-up" scene analysis to identify such features. From a systems point of view, Pribram's (1971) metaphor, based on holographic processing by excitation pattern interference in a parallel processing neural network, proves such "top-down" attunement to be perfectly possible.

The Internal Dynamic World Model

The dynamic attunement of the organism from this point of view clearly plays a very central role. The existence of such an internal world model cannot be doubted, only the research model used to represent this attunement can be subject to discussion, as well as the segregation of the function as a separately existing dynamic model. A kind of dynamic world model is necessary in order to account for control of responses to the environment that are too fast to allow control by simple perceptual feedback. Often-cited examples are fast sequences in sport, musical performance, etc., and the quick draw of western gunmen, which has recently been used for military training (Arnold, 1969). To serve this purpose, it is necessary that the internal dynamic world model simulates not only the behavior of the environment, but also of the body; i.e., it simulates the interaction. The dynamic world model also manifests itself when humans respond to events that do not happen or information that does not appear in a familiar context. Operators in industrial process plants have a process feel that enables feed-forward control of plants within a certain span of time constants (Crossman and Cooke, 1962), and they may have complex responses which can be released by very simple cues, for example, auditory warning signals.

The existence of an internal world model has important implications. An operator cannot be considered merely as a data channel transforming input information into actions. Instead, in a specific situation, the input information synchronizes the internal model, and complex responses may be generated from primitive inputs. In highly trained situations, it can be extremely difficult to predict the information that is used to synchronize the internal model. Often it can be secondary sources, utilizing such auditory information as relay clicks and motor noises. Warning signals in process plant control rooms are generally meant by the designer to be signals to alert the operator, who is then supposed to consult the instruments to identify the cause. In case of rather frequent warning signals that appear as a normal part of a work sequence, during plant start-up for instance, the internal model of the operator apparently gives him a "process feel," which continuously keeps him "set" for the proper action. He often only needs to be triggered to take action. He "knows" from the general behavior of the system and from his prior actions what has to be done without consulting the instruments. He may even–without realizing it–correctly respond differently to the same warning signal in different operational phases–for instance, during start-up when the properties of a plant may change through time.

The time span covered by the dynamic world model is demonstrated by the experience which Pribram (1969) calls the "Bowery-El Phenomenon":

> ... For many years there was an elevated railway line (the "el") on Third Avenue in New York that made a fearful racket; when it was torn down, people who had been living in apartments along the line awakened periodically out

> of a sound sleep to call the police about some strange occurrence they could not properly define. Many such calls came at the times the trains had formerly rumbled past. The strange occurrences were of course the deafening silence that had replaced the expected noise...

This example also illustrates another important function, a mismatch detection that alerts conscious control, if the state of affairs in the environment does not behave according to the state of the internal model. This occurs when the internal model loses synchronism or is not properly updated due to unexpected behavior in the environment.

The Output System: Motor Functions

The output from the subconscious data-processing system is always a motor function, a movement, or a speech act. The motor system consists of a great number of muscles that are controlled and coordinated by a complex, highly interconnected nerve system. It is generally accepted that this coordination and control is hierarchically organized, and experiments show that the modeling function of the subconscious data processing is also distributed in the different levels of the motor system. This makes the motor system of the body an active part of the human information processor.

The closely integrated nature of the entire system is also demonstrated by the way the internal model and the generic features build up. The model and the generalization are not the result of passive processing of the perceived information, but a result of active, physical interplay with the environment. This statement is supported by experiments; see for instance, the experiments by Held and Hein (1963). These aspects are not only of academic interest but may have important practical implications. The movements of the head that are needed to read instruments in a control room may take part in the identification of the source of the information. The physically well-defined data, such as distance or speed, which are used as discrete data for inputs in a mathematical treatment of a dynamic process, are not defined or used by the subconscious data processing as separate parameters, but implicitly as part of a pattern having a functional meaning. For instance, distance is not judged or measured by a two-eye stereoscopic vision, but by the operational consequence experienced by walking or reaching along it.

Like the perceptual function, the motor control is not based on stored behavior patterns from prior encounters, but on a constructive process that generates the proper patterns on demand (Bernstein, 1967). This is demonstrated by the fact that the success of rapid movements is independent of the initial positions of limbs and that movements can be transferred to other metric proportions and limbs. This function must depend on schemata for generating complex movements with reference to the internal dynamic world model. An important ingredient in motor control is the dynamic feed-forward generation of patterns within this internal dynamic map, which is updated and aligned by the sensory information.

Conscious Data Processing

Conscious control of the direct interaction with the environment is only necessary when a mismatch with the behavior of the environment results in an interrupt signal calling for immediate attention. Thus the high capacity of the subconscious sensorimotor functions protects the low capacity of the higher-level conscious cognitive functions from overload in familiar routine tasks. Conscious attention may be considered as a single-channel function that has to be switched between different items, objects, or tasks. Therefore, the conscious data processor may be thought of as a sequential processor that normally runs different tasks on a time-sharing basis. The limiting properties of the conscious data processor are related to the capacity of its short-term buffer memory, which is generally taken, as a rule of thumb, to be between two and seven items of information (Miller, 1956; Simon, 1969). The effectiveness of the conscious processor, in spite of these limitations, is due to both its ability to operate on efficient information codes at high levels of abstraction and to its large repertoire of different data-processing models and strategies. The role of the conscious processor in overall performance varies widely. It can be used for passive monitoring of the performance of the subconscious processor in routine tasks. When initiated by interrupts caused by less familiar situations, it can be used to bridge mismatches of the subconscious processor by categorizing the input set and adding verbal labels to situations, thereby switching states of the subconscious processor. It can also perform problem solving in unique situations by evaluating alternatives and by making decisions and plans based on predictions, etc. During problem solving, the conscious data processor is capable of operating in several basically different modes, which implies cooperation with the subconscious processor at different levels. For example, it can simulate the external world in the domain of the sensed information (e.g., make visual experiments in which a visual imagination of a process is used to foresee the response of the environment to planned manipulations), or it can process symbolic data by following a prescribed plan, a sequential procedure (e.g., to make numerical calculations and abstract logical reasoning).

To be able to interact successfully with a physical system, the state of the system and its response to the actions intended have to be identified from the observations or data received from the system. Generally, the observations have no meaning individually, only the interrelationships in a set of data will define the state of the system and allow a prediction of future behavior. This interrelationship is due to the constraints imposed upon the set by the internal anatomy and processes of the system. The human data processing has to be based upon a representation of such system-given constraints. This representation is a data-processing or transformation model of the system and can be present as an internal representation (i.e., a mental model), or external models (i.e., drawings, diagrams, etc.) can be used to support the data processing. The mental model is not necessarily a visualization of the physical system, but can be a more abstract data-processing model in the engineering sense. Such models have already been discussed by Craik (1943). Furthermore, a procedure is needed to use the model for data processing. The conscious data processor has several different models and procedures at its disposal, and the choice in a specific task depends upon different characteristics of the tasks.

Examples

The problem is to judge the movement of a mechanical lever when another lever is operated by hand (Sloman, 1971). The levers are connected internally in the system by a mechanism of ropes and pulleys. This problem may be solved by an internal visualization of the system or by a simple sketch on paper. Some observers may be experienced enough to grasp the solution as a whole while looking at the sketch; others may have to proceed through a series of intermediate steps–if I move this lever this way, then..., and so forth. The procedure will then be a stepwise tracing through the system of the flow of movements, a stepwise visualization of the effects of a simulation. The model as well as the procedure used has a close relation to the actual physical problem. This sort of direct visual experiment is possible if the physical system and the relevant variables are directly visible; if not, some kind of visual analogy can be used. For instance, this is done when a primary school teacher uses "water-pipe-and-jar-explanations" to explain simple electrical circuits and when mechanical mass-and-spring systems are used to visualize electrical resonance phenomena. In this type of data handling, the model supplying the system constraints between the individual observations is either the physical system itself or some kind of physical or visual analogy, and the procedure to obtain the desired resulting data is a visual experiment, i.e., you visualize the process itself.

If the process or the system is too complicated, other types of models are generally used. A model of the constraints often used in a design or research task is the abstract or symbolic model constituted by mathematical equations. A system of equations models the constraints between the individual variables or observations. The characteristic feature of such a model is that it is generalized. The data-handling problem has been transferred to a symbolic world where neither the model itself nor the procedure by which it is used now has a clear relation to the physical world. Several procedures are available and used to process data by equations. One is a sequence of logical, symbolic decisions and manipulations based upon a set of mathematical theorems. The written equations are then a visual support for the symbolic mental operation on numbers, sets, and relations. But the written equations live their own lives, and the procedures used for treating familiar types of equations look more like direct visual manipulation of the physical signs on the paper. You know how they behave when you move them around. Signs change when symbols are transferred from one side of the equations to the other; $d(x^2)/dx$ changes almost by itself to $2x$, and so forth. The display aspect of solving algebraic equations has been discussed by Myron Goldstein (1969). The power of the mathematical data processing is due to the generality of the model and existence of a comprehensive system of established procedures, which are only related to the formal model itself, not to the physical problem. Other mental models related to the physical world are of course needed to transform the problem to the formal domain. However, they do not necessarily represent the total system, but only the behavior of the parts at the level where the individual equations are formed.

Complex data processing by mathematical models and procedures is typically used by system planners and designers rather than system users and operators. If the data processing implied in system operation in a certain situation calls for a complex data handling procedure–e.g., solving of mathematical equations–then it has very reasonably been foreseen by the system designer. He will then perform the data processing necessary to decide upon the appropriate sequence of manipulations or actions by means of experiments in the laboratory, by simulation using some sort of analogy–physical or mental–or by use of mathematical equations. From the results of this data processing he can then extract a set of operating instructions or operational procedures to be used by the system operator. The data handling procedures of the operator are now simple: to execute a stored sequence of predetermined actions. The system model supporting the data processing now has to supply the operator with state-defining data sets, to initiate the stored program, and to provide a description correlating the prescribed actions with the manipulation surface of the system, i.e., some sort of topographic map. The functional model of the interior of the system is now found implicitly in the stored program. Whatever the nature of the transformation model that has been in initial system operation, experienced system operators develop work procedures or rules by storing sets of state-defining observations and goal-directed procedures in a kind of decision matrix, as discussed by Beishon (1966). Even when operating instructions have been issued by the system designer, system operators often tend to form their work procedures directly by an initial trial-and-error operation–i.e., to derive the transformation model needed directly from the system behavior–rather than spend the effort reading the manual.

These examples represent basically different ways to control the overall behavior, which will be discussed more formally in Chapter 9.

The information necessary for the various categories of conscious processing will be sampled from the environment by the perceptive front-end system. This means that even conscious information processing depends on a properly updated and synchronized world model to give the basic context, i.e., to guide observation and define the proper level of probing invariant features. Or, the other way round, the extent to which the subconscious dynamic world model includes the behavioral properties of elements in the environment, which may be symbolic representations on paper, determines the level and efficiency of the conscious strategies that are available.

There is an important difference between the internal models or representations of the environment that are used by subconscious sensorimotor responses and those of the higher-level cognitive processes, a difference that has a fundamental influence upon the kind of research model that can be used to represent operator behavior. Subconscious, sensorimotor responses appear to depend upon an active, dynamic model that simulates the behavior of the environment. This means that the data processes are determined by the structure and elements of the models, conceptually in the same way as the processes of an analog computer are determined by the laws of nature, not process rules, when it is initialized and activated. In contrast, the sequential, conscious data processes are typically based on stationary (static) mental models or representations of the functional properties of the system to be controlled, in the form of cognitive, functional maps. Process rules or strategies are therefore necessary to activate and control the steps in a sequential data process.

In relation to the cooperation between the conscious and the subconscious processing, it may be illustrative to consider the recent development of the theories of imagery, which relates problem solving to the functions of perception and the internal world model.

Imagery and Problem Solving

Visualization and imagery are having a renaissance as acceptable concepts involved in human information processing. After having been discarded for a long period during behavioristic dominance in psychology, the work of Gibson (1966), Neisser (1972), and others has again made it acceptable to consider imagery seriously. However, a change has taken place from considering imagery as stored traces in terms of mental pictures from previously experienced sense data toward the view of imagery–and memory–as a more constructive process depending upon the "attunement" of the neural system to invariant structures in the array of sense data. Gibson (1966, p. 277) reacts to what he calls the "store house" theory of memory, and Neisser (1972) continues: "What is relevant is the notion that one normally perceives layouts, not pictures. If this is true, we should stop talking about images as mental pictures and try to understand what a mental layout may be." This brings the arguments close to the view that the interactions of the body with a dynamic environment as well as manipulation of objects and tools (and symbols) depend on an internal, dynamic world model that has many features in common with a versatile analog model. A flexible analog model can be structured from prototypical representations of objects and environments and synchronized by "resonance" with invariant features in the environment.

Because attention and memory–understood as attunement–depend upon the function of the dynamic world model, it is to be expected that this model has a major role to play in higher-level cognitive functions such as reasoning and problem solving as, for instance, based on imagery and visual thinking. This, however, does not necessarily mean that introspection and imagery reveal the mental processes. Thinking may heavily depend on subconscious processes related to the structure of the model evoked, whereas imagery and verbal reports are only the manifestations of the coupling of the processes to the conscious attention and control. Imagery makes the structure or mental "layout" available to conscious attention, and verbal expressions may be applied to control or report on changes of the state of the model.

If both imagery and perception depend on the actual structure and content of the internal, dynamic world model, various kinds of interferences should occur. This has been studied experimentally by Brooks (1967, 1968) and Segal (1971), who showed that visual imagery suffers when it is put into direct competition with visual perception, while auditory/verbal imagery suffers from competition with listening and speaking. They also found poorer performance in recall of sentences, which referred to spatial layout, when sentences had to be read, than when they were presented orally. A remark in passing: it should be noted that it must be considered whether "mental layouts" can be found at higher levels of abstraction than the visual representation of the concrete physical world. Michotte's (1963) experiments of direct perception of causality seem to indicate such higher-level mental layouts. And it should be considered whether information processing at such levels can be effectively supported by forming "concrete symbols" (graphs, drawings), i.e., by defining temporal-spatial symbolic representations. Also, the danger of interference discussed by Brooks should be carefully considered when graphic (i.e., spatial-temporal) or alphanumeric (i.e., verbal) displays are used to support information processing.

Recently, several authors have argued that imagery and verbal formulations have different roles in information processing and problem solving. (A review is given by Kaufmann, 1979.) Horowitz (1967, 1970) regards thought as a process that involves multiple cognitive systems, including enactive, iconic, and lexical, where iconic includes imagery and lexical involves verbal expressions. Atwoods (1971) studied interference between learning based on imagery and verbal learning and concluded that imagery and verbal symbolic processes represent alternative coding systems in cognition. Paivio (1971a,b) has studied imagery. He conceives imagery to be related to concrete aspects of a situation, with the verbal processes to be more functional as mediators in abstract situations. Imagery is considered to be dynamic, to promote flexibility and transformational processes, whereas verbal functions are considered more static, labeling functions. He also characterizes imagery as visual, spatial representations suited for parallel processing, whereas verbal processes are specialized for sequential processing. This distinction appears to be too simplistic. Imagery may be suited for parallel processing of variables in the sense data, but verbal processes can deal with sequences of very complex state representations. This aspect will be discussed in more detail later. Paivio also argues, as has been done above, that imagery need not be available to consciousness to be functional. The spatial structure makes imagery very well suited to represent an overall problem situation, to give an overview that serves to structure or restructure a problem in accordance with previous experience, and to identify invariants in the same way as Gibson's direct perception. The role of imagery in early phases of problem solving when simultaneous consideration of several aspects is necessary has been mentioned by several authors.

The French mathematician Hadamard (1945) resorted to imagery in his creative efforts when matters grew complex "in order to have a simultaneous view of all arguments, to hold them together to make a whole of them." Rugg (1963) considers two phases in creative thinking. The first phase–discovery–involves a transformation phase by imagery, while the second phase–verification–is presumed to involve the more logical and direct verbal system. However, the process of this initial discovery phase may not itself be imagery; the initial phase may be due to subconscious processes related to the attunement of the neural system, i.e., to the internal dynamic world model. Imagery may be a possible, but not necessary, conscious manifestation of this process. However, because it will even then be closely linked to the mechanisms behind perception, the mechanisms of interference discussed above must still be taken into account, for instance, in display designs.

The intuitive phase of problem solving involving subconscious, holistic processes has been mentioned by several authors, frequently in relation to imagery. Bruner et al. (1966, p. 57) distinguish between analytic and intuitive thinking:

> Analytic thinking proceeds a step at a time. Steps are explicit and can usually be adequately reported by the thinker to another individual. Such thinking proceeds with relatively full awareness of the information and the operations involved. It may involve careful and deductive reasoning, often using mathematics or logic and an explicit plan of attack. Or it may involve a step-by-step process of induction and experiment. Intuitive thinking characteristically does not advance in careful, well-planned stages. Indeed, it tends to involve maneuvers based seemingly on our implicit perception of the total problem. The thinker arrives at an answer, which may be right or wrong, with little awareness of the process by which he reached it.

This description matches very well the situation when an analog representation simulated a scenario without the need for "process rules."

From this discussion it appears that intuitive thinking related to mechanisms of perception and imagery is important for the initial problem formulation or restructuring phase, for transformational purposes; whereas thinking related to verbal symbolic processes is predominant in rational, sequential processes of verification and planning. Berlyne (1965) argues that imagery, conceptualized as internalized action, is of special importance in transformational thought, in contrast to language, which is supposed to operate in situational thought, to have primarily labeling functions in thinking. Lately, Pylyshyn (1973, p. 10) has argued that images may be especially useful in the process of creating new information: "While picture-like entities are not stored in memory, they can be constructed during processing, used to make new interpretations and then discarded."

In general, there seems to be too sharp a distinction between the role of imagery and verbal symbolic processes, between intuitive and rational thinking. Probably, both categories of processes play an important role in all thinking and problem solving. From an information processing point of view, the basic process in conscious problem solving can very likely be subconscious symbol manipulation or transformation based on processes related to the "attunement" behind perception. This process only appears as imagery if the integrated flow of transformation stops, and the individual consciously starts a search to restructure the problem situation by considering the symbolic mental layout that may appear as "imagery." Verbal processes are used to activate or to label the state of the subconscious attunement behind skilled symbol manipulation and to consciously control a sequence of such processes when a familiar problem-solving sequence is found applicable.

A key issue appears to be the relation between language and the mental processes behind imagery and perception. A conceptualization of memory organization, which is very compatible with the above view of the internal world model as an analog representation within a parallel-processing neural network, has been suggested by Norman and Bobrow (1975). They propose a network of memory schemata with the following special features. They are active and can communicate mutually—which are typical features of the elements of a simulating, parallel-processing network. They are activated by "context-dependent descriptions," and the concept leads to a process of perception that depends on data-driven, bottom-up processes, as well as concept-driven top-down processes, which matches the model of Figure 8.6. Also, the role of the active memory schemata in the subconscious problem solving as discussed by Norman and Bobrow is basically similar to the view presented above.

The view that verbal processes serve to label situations and states in the environment and to formulate intentions, and thereby to update or modify the current internal dynamic world model, is compatible with the results of the work of Vigotskii, Sokolov, and others; see Sokolov (1971). These studies analyze the cognitive performance of persons with hindrance of internal speech by secondary verbal tasks or with blocked speech organs, in order to find the role of internal speech in mental processes. The general finding is that speech hindrances greatly impede the comprehension of texts read to persons and their problem solving, as, for instance, the solution of arithmetic problems. If internal speech was hindered by a secondary verbal task, the quality of the primary task improved as the secondary task became automated, and the verbal statements of the persons indicated that they were now able to insert rapid key words articulated soundlessly and thereby to control the primary task. Also, they reported the use of visualizations as mnemonic signs.

Visual images combined with reduced verbalization seem to be a frequently observed phenomenon in such experiments. The internal speech in this situation is

> ... no longer merely soundless speech, but presents a considerably transformed kind of the latter, especially adapted to the performance of the thinking functions. In essence, that which is fixed in internal speech presents only a "thinking model" or a "thinking plan" of the speech actions, which in some cases may be greatly contracted like a telegraphic code, while in other cases it may be more extended, turning into an "annotation of a statement" or into an "internal monologue." Strictly speaking, the first, reduced form of internal speech does not contain any words in their usual grammatical meaning; the words are rather implied by being replaced with their kinesthetic code, which only reminds one of an articulation of the key words. Numerous examples of such speech reductions are cited by Vygotsky (1934); he connected them with the semantic predicativity of internal speech, with the predominance of predicates and the lack of subjects.... [Sokolov, 1971, p. 86.]

This is precisely what is needed to update or "strobe" a person's own internal world model, in contrast to the more fully developed verbal expression needed when communicating in order to activate a particular world model in another person. Similarly degraded verbal expressions can be found in conversation between two persons deeply involved in the same work situation and therefore having highly synchronized world models. Sokolov also analyzed internal speech in problem-solving situations and found, by measuring electromyographic potential, that speech motor excitation increased with complexity and novelty of a mental task, for instance, when a transition occurred from one type of arithmetic problem to a different type. Again, the verbal expressions used by subjects were of the reduced form that constitutes one of the characteristic features of visual thinking.

> The situation of such thinking does not require any verbalization of all that is perceived; the internal speech functions in a highly generalized and fragmentary way, only directing the process of visual analysis and synthesis and introducing certain correctives into them. [Sokolov, 1971, p. 90.]

Or, in other words, the verbal statements only serve to direct the state of the internal world model underlying thinking and perception.

Dynamic World Model and Mental Models

In modeling human performance, it appears to be useful to distinguish between the dynamic representation underlying automated sensorimotor performance, and the more static mental models used for conscious, sequential reasoning. However, there is a tight relation between these types of models, which will be particularly clear when considering visual imagery used for conscious problem solving.

The dynamic world model of automated sensorimotor behavior can be seen as a structure of hierarchically defined objects and their behavior in a variety of familiar scenarios, i.e., their functional properties, what they can be used for, and potential for interaction, or what can be done to them. These elements of a generic analog simulation of the behavior of the environment are updated and aligned according to the sensory information during interaction with an environment. The model is structured with reference to the space in which the person acts and is controlled by direct perception of the features of relevance to the person's immediate needs and goals.

In many ways, this conception has similarities to Minsky's (1975) "frames." The main–and fundamental–difference is that Minsky's frames depend on a sequential scene analysis; they are structured as networks of nodes and relations, and they are basically static. Minsky defines frames as a data structure for representing stereotyped situations that are organized as a network of nodes and relations. The top levels of a frame are fixed and represent things that are always true about the supposed situation. The lower levels have many terminals that must be filled by specific instances or data. This means, however, that the overall spatial relation will be replaced by local relations as "left-off" or "above." The dynamic properties are replaced by "before-after" frame pairs. In a way, Minsky's frames look like a sequential, introspective analysis of the appearance of the subconscious dynamic world model during visual thinking and imagery. Minsky's use of one single concept–the frame system–to model the human representations of the environment that are active for sensory perception, for physical and verbal interactions with environment, and for symbolic reasoning, leads to unnecessary simplification. It appears to be more fruitful to rely on various types of representations. Gibson's (1966) concepts related to "direct perception" are far more convincing, viewed as a model of the high-capacity information-processing mechanisms underlying perception, sensorimotor performance in fast sequences, etc., than Minsky's symbolic information processing. The latter is more adequate for higher-level conscious information processing, i.e., the manifestations of the "dynamic world model" at the conscious level in terms of natural language representations.

As I understand Gibson's concept of direct perception, the "dynamic world model" is in the present context very similar to the mechanisms needed for the "attunement of the whole retino-neuro-muscular system to invariant information" (Gibson, 1966, p. 262), which leads to the situation where "the centers of the nervous system, including the brain, resonate to information." This selective resonance relies on the existence of a generic dynamic model of the environment. The implications of Gibson's view of perception, as based on information pickup instead of sensation input, are in many ways compatible with the model of human information processing illustrated in Figure 8.6. To Gibson, perception is not based on processing of information contained in an array of sense data. Instead, the perceiver, being attuned to invariant information in space and time in the environment, samples this invariant information directly by means of all senses. That is, arrays of sense data are not stored or remembered. They have never been received; instead the nerve system "resonates." In my terms, the world model, activated by the needs and goals of the individual, is updated and aligned by generic patterns in the sensed information, but the idea of an organism "tuning in" on generic time-space properties is basically similar and leads to the view of humans as selective and active seekers of information at a high level of invariance in the environmental context. The subconscious dynamic world model or the attunement of the neural system leads to the situation where primitive sense data are not processed or integrated by symbolic information processes as Minsky suggests, but the generic patterns in the array of data in the environment are sampled directly by high-level questions controlling the exploratory interaction involving all senses.

A related point of view is argued by Rosch (1977, p. 30, cited from Dreyfus, 1981): "Many experiments have shown that categories appear to be coded in the mind neither by means of lists of each individual member of the category, nor by means of a list of formal criteria necessary and sufficient for category membership, but, rather, in terms of a prototype of a typical member." Such prototypes will be similar to generic elements in an analog representation of the environment. Rosch continues to make the correspondence even clearer: "The most cognitively economical code for a category is, in fact, a concrete image of the average category member." As Dreyfus adds: "One paradigm, it seems, is worth a thousand rules" (1981, p. 178).

Gibson draws some general conclusions on memory and learning that are related to the role of the "dynamic world model" at the subconscious level of Figure 8.6. Regarding memory, he states, as already cited in relation to perception (Gibson, 1966, p. 262): "It should now be clear that the brain does not have to integrate successive visual sensations in immediate memory. There is no necessary reason to suppose that the fixations have been retained. The invariance of perception with varying samples of overlapping stimulation may be accounted for by invariant information and an attunement of the whole retino-neuro-muscular system to invariant information. The development of this attunement, or the education of attention, depends on past experience but not on the *storage* of past experience." This conforms closely to observations of behavior in process control, where data are not remembered, but their functional implications are.

The states of the internal world model or the attunement to the environment of the neural system are reflected in higher-level information processes in terms of natural language. Verbal statements reflect perceptual structures, and can be used to report perceptions, to activate world models for imagery, and to synchronize the world models of several persons through verbal communication. Sequential scene analysis and verbal reasoning based alone on semantic nets or "frame systems" defined as a system of nodes and relations will lack the basis for the overall "direction" in the process. It will be like walking in a city without a map, only aware of the immediate neighborhood. Only by assuming the attunement of the system or an effective world model will there be a basis for overall control of the direction in the sequential process.

From the beginning, the AI community seems to have been aware of this difficulty with a frame concept in terms of a network of nodes and relations. Goldstein and Papert (1977) introduced anthropomorphisms like "frame keepers" and "demons" for procedural control, and state (p. 93):

> In the previous paragraph, we wrote as if the tickle-frame could be *direct, explicit* knowledge about feathers. A fundamental and pervasive class of problems concerns the relative feasibility of achieving this as compared with various schemes for *indirect* reference to feathers via some *superset* or *description.* For example, the tickle frame might know that "long, soft objects" are good instruments for tickling and somehow use this as a means for finding a connection to feathers. Most early workers in AI (like most contemporary psychologists) were strongly averse to the direct approach, and so a great deal of attention has been given to the invention of "search" and "information retrieval" methods to mediate the indirect approach. Current thinking (especially among frame theorists) leans towards very much more direct knowledge than was previously considered feasible. This places a heavy demand on memory, and less on retrieval mechanisms. This appears to some to be "heavy-handed." But we do know that human memory has a very large capacity, and no one has been able to propose very sophisticated retrieval mechanisms. So perhaps it is nature (rather than the theorists) who has adopted the "heavy-handed" solution.

For perception and sensorimotor activities, Gibson's attunement appears to be a much more effective and elegant solution than this "heavy-handed" concept, and even for sequential verbal reasoning, organization of the content of mental models in terms of schemes for "indirect reference to feathers via some superset" elaborated to a multilevel abstraction hierarchy will be more acceptable to account for the direction in reasoning than "direct, explicit knowledge." Duncker (1945) found that problem solving is a sequence of reformulations of a problem, "of mediating phases" of which each one, in retrospect, possesses the character of a solution and, in prospect, that of a problem. This reformulation considers several levels of "functional values," their "by-means-of-which" features, and Duncker remarks: "There is no need for a special concept of a "direction" that combines elements. Direction is of the type of a *problem,* or, more exactly, of the reformulation of a problem and of a mediating phase in the solution process." This view is entirely depending upon a process through several levels of "functional values." Functional value is tightly related to somebody's intention or reasons, as shown in Figure 4.1. A chair is only useful as a chair if somebody wants to sit down; and if so, many objects may have functional value for sitting, if a chair is not present. The form and content of mental models will be discussed in more detail in Chapter 10.

The Internal World Model Revisited

It is clearly necessary to have a more specific characterization of the internal dynamic world model or attunement of the organism. The view advocated here is compatible with the positions of Gibson (1966) and Pylyshyn (1973) that knowledge of the environment is not stored in memory as traces of sense data, but as highly processed information. This makes it possible to regenerate imagery and verbal descriptions that are only to be taken as the manifestations of the stored information structures in the conscious domains and the means for conscious control of the knowledge representations. From the role in human behavior, some of the functional features of the world model can be summarized:

It is able to control bodily movements in a feed-forward mode of control during fast sequences, i.e., it is capable of real-time, quantitative, and precise simulation of the time-space patterns of the environment, and it is an *active* model.

It is a *hierarchical* representation; it enables recognition of objects and scenes at the level of physical appearance; it makes it possible to identify objects by their functional values rather than their appearance; and patterns of purposive behavior can be activated by high-level intentions.

There is a very efficient mapping between features of the environment and the model; i.e., a very efficient updating of the model is possible in response to changes, as well as easy transfer to "similar" scenarios. This points to an analog-type model with elements representing objects and their functional properties and values, and consequently a one-to-one mapping of elements in the environment and the model.

Pylyshyn (1973) discusses the need for knowledge representations of a more fundamental nature than imagery. In accordance with the view put forward here, he points out that one has to accept imagery as a phenomenon of importance for cognitive behavior but that imagery does not reveal the information processes. Instead of accepting the process of perceptive analysis of imagery as the basis of visual thinking, the visual imagery should rather be seen as the result of underlying perception-like processes. He finds the typical features of imagery "much closer to a description of the scene than a picture of it" (p. 11) and therefore it is "useful to think of propositional knowledge even when the concepts and predicates in such propositions do not correspond to available words in our vocabulary" (p. 7). Consequently, he discusses as models of knowledge representations the propositional representations, data structures, and procedural representations developed within computer science. Like Minsky's frames, these representations have great advantages by being immediately operational for digital computer simulations. However, these kinds of Fregean knowledge representations have some shortcomings when taken as psychological theories.

First of all, they tend to be passive representations, needing external control algorithms; for propositional representations in terms of rules for deductive reasoning, for data structures in terms of retrieval routines, which lead to the procedural models that need still higher-level rules for proper functioning, they all lack the active, reconstructing power of the representation behind perception and imagery, and, just as for Minsky's frames, it is tempting to invent a "demon" for this control (cf. Norman's plea for active memory units).

Second, among abstract concepts, i.e., attributes of elements in the environment, the Fregean character in consequence leads to a complex mapping function between changes in the environment and necessary updating of the representation. It appears plausible to relax a little from digital computer biases and consider the spatial characteristics of the neural basis for representation. The neural networks are basically capable of active representation or simulation of properties in the time-space patterns of information from the sensory data.

Analog processes representing the behavior of the environment are possible in a pulse-density-coded network seen as a distributed digital differential analyzer (Ribeiro, 1967) or as a holographic processor (Pribram, 1966). By holographic processes, the neural network would be able to generalize at several levels of abstraction and to identify "prototypical" objects, functional properties, and value features given the activation by suitable "reference beams." Analog applications of holographic techniques have been demonstrated in optical data processing in terms of optical Fourier transformation for object and feature identification from prototypical reference samples. The point in question is that search for a model of the basic knowledge representation should not be constrained by the bias from symbolic processes in sequential digital computers, but should include the possibility of a much more effective mapping of active features in the space-time information contained in the sensory data. Holographic generalization from the sensory time-space array of information by reference beams related to functions, goals, and purposes may very well lead to a hierarchical and prototypical analog model up through a hierarchy of abstraction similar to the one shown in Figure 4.1.

A holistic, holographic process based on stored, dynamic interference or correlation patterns between excitation wavefronts seems to be very plausible. A hierarchical organization can be based on identification of generic spatial patterns representing objects, i.e., invariant over scenes and during movement and related to human acts. The next level of functional properties can be identified by invariants in events of object interaction and value properties by invariants in relationships between object patterns and human intentions. Michotte's (1963) experiments with perception directly in terms of causal relation between objects, when subjects were presented to abstract dynamic light patterns, seem to indicate such mental "layouts" at a higher abstract, functional level. Furthermore, based on Johannson's (1973) experiments with perception of human movements, when persons with small light bulbs attached to the major joints moved in the dark, Annett (1981) asserted that a human's perception of movement is tuned for reading intentions, i.e., for immediate perception of what other people are about to do. Holographic memories depend on "reference beams" for retrieval of information. Such "reference beams" can probably be generated in the neural network by excitation waves from other neural complexes; intentions may activate reference patterns from the speech motor system, which in turn activate the holographic representation top-down, creating imagery. Sensory output can activate bottom-up and lead to perceptions. This holographic conception very well matches the requirements of Pylyshyn (1973, p. 10) that vague recollections should be characterized by lack of resolution, or "in the sense that certain perceptual qualities or attributes are absent or uncertain–not that there are geometrically definable pieces of a picture missing." (If part of a hologram is removed, the entire content will be developed by a reference beam, but with decreased resolution.) This active knowledge representation will be able to control bodily interaction with the environment as well as visualization verbalization during all routine situations. During unique situations, the simulation may stop, and outside control may be needed. If the motor schema of speech is an independent coding form, it will be plausible that internal speech will be effective to control the dual process of detailed operation on the knowledge representation manifesting itself as imagery, and keeping track of the overall plan by means of speech motor schemata.

The difficulties with an entirely Fregean or propositional model of knowledge representation have recently been discussed by several authors. Sloman (1971) argues the need for a more analog and intuitive model of reasoning to represent, for instance, the immediate "grasping" of the function of a pictorial presentation of a pulley. Pylyshyn (1973) realizes the difficulties with the propositional models of knowledge representation as a model of human cognition (p. 13). The system must, for instance, "display similar *intermediate states of knowledge* as subjects do in solving a problem or answering a question.... It is not clear at this stage in our understanding of theorem-proving schemes whether a uniform proof procedure is capable of meeting such requirements." This condition will serve as a test at only the higher symbolic level of the information processes, not as a test of the model of the more fundamental aspects of knowledge representation at the level of processes behind imagery. A more reliable test of a model will be the requirement that it must make mistakes and errors in the same way and in the same situations as humans will do. Johnson-Laird (1980) discusses in length the problems related to a propositional model of knowledge. The problems arise with inferences such as: A is to the right of B; B is to the right of C; then A is to the right of C only if on a straight line. If sitting around a round table? The problem is that the concepts of right and left are only defined in a given structural context. This problem is not in particular related to right and left, but many terms are ambiguous unless the structural context is specified separately: "at," "between," "near," "next to," etc. To generate propositions is to abstract selectively some properties from the context and, consequently, the structural context cannot be regenerated from the propositions. This makes a propositional theory of imagery difficult because "it must be possible to apply the inverse of the propositional encoding to obtain the original stimulus, or, more plausibly, a sensory representation isomorphic to the original stimulus. However, because perception is likely to involve a many-to-one mapping, the inverse may fail to yield the original 'stimulus'" (Johnson-Laird, 1980, p. 94). To resolve such difficulties, Johnson-Laird introduces the "mental model" which, however, seems to be a model only at the level of representing the spatial relationships among objects that are labeled by names, categories, and relational properties. The possible extension of the concept of mental model to more structural, symbolic levels is not discussed by Johnson-Laird.

In conclusion, the most economic and effective representation of knowledge appears to be based on an interaction between a dynamic world model representing the spatial structure of the state of affairs of the physical world directly, as well as at different symbolic levels, and a system of verbally coded propositions serving to control the simulation processes. The relationship among the concepts, which is the basis for the propositions, the knowledge base, is a "mental model" of the environment.

This mental model or cognitive map can be viewed as a network of relations among conceptual nodes, a "semantic net." However, the interpretation of the verbal codes of this model heavily depends on the context supplied by the current, subconscious dynamic world model. Attempts to define the meaning of isolated words and even sentences are only meaningful within formal logic, not in natural language.

The need for a function like the internal dynamic world model as the background for understanding language, as well as for actions, has long been discussed in the philosophical literature. Polanyi (1967) has thoroughly discussed the importance of "tacit" knowledge. Machie (1975) found it necessary to introduce the notion of a "field" as the representation of the context in his efforts to define causality in commonsense descriptions of event sequences. Recently, Searle (1983) has argued the importance of the "background," which in his terms is the "nonrepresentational" something underlying mental representations, "intentional states." Mental representations form a network of intentional states, and the semantic content of a state depends on its location in this network. However, "anyone who tries seriously to follow out the threads in the network will eventually reach a bedrock of mental capacities that do not themselves consist of intentional states (representations), but nonetheless form the preconditions for the functioning of intentional states." His arguments for the existence of the "background" is very analogous to the present arguments for the internal world model: the background is necessary to account for the fact that the literal meaning of a sentence is not a context-free notion, for understanding of metaphors, and to explain physical skills as, for instance, needed in expert skiing.

The speculations above are by no means intended to serve as a hypothesis on the actual organization of human information processing. Instead, they may serve to illustrate a possible organization as a counterexample, considering the trend in cognitive psychology to take the AI conceptualizations as psychological theories. It is evident that concepts within AI must be related to the means available for implementation today. It should not, however, prevent the consideration of other models by the system designers, if they give a more realistic representation of human capabilities and limitations, even these cannot be reproduced today by existing technology and thus be tested by direct simulation.

Fixation on current sequential, propositional AI models may very well prevent a designer of human-machine interface systems from considering possible alternative and efficient, high capacity human processing modes.

Chapter 9

Models of Human Information Processing

In order to have a basis for man-machine systems design, it is important to identify manageable categories of human information processes at a level that is independent of the underlying psychological mechanisms. Such implementation-independent models furthermore will be compatible with models used for design of the automatic information-processing algorithms of the interface.

Distinction of various categories of human information processes appears when we consider the way in which human behavior can be internally controlled. This leads to a more formal description of the various domains of behavior behind the examples discussed in Chapter 8.

In this effort, we have to consider that humans are not simply deterministic input-output devices, but goal-oriented creatures who actively select their goals and seek the relevant information. The behavior of humans is teleologic by nature. In their classic paper, Rosenbluth et al. (1943) define teleologic behavior as behavior that is modified during its course by signals from the goal. This restrictive definition seems, however, to be due to an inadequate distinction between two concepts: *causes* of physical events and *reasons* for physical functions, a distinction that has been discussed in detail by Polanyi (1958). Teleologic behavior is not necessarily dependent on feedback during its course but on the experience from previous attempts, i.e., the reason for choosing the particular approach. Reasons act as the classic "final causes" and can control functions of behaving systems by selection, be it natural selection in biologic evolution or through human design choices for man-made systems, whereas causes control functions through the physical structure of the system. In that all technical systems are designed for very definite reasons, it directly follows that teleologic explanations–in the classical sense–of the functions of man-made systems derived from their ultimate purpose are as important as causal explanations based on engineering analysis. The same is the case for explanations of purposive human behavior.

Actually, even human position and movement in the physical environment are only occasionally directly controlled during the course of action by simple feedback. It may be the case in unfamiliar situations calling for accurate and slow time-space coordination, but in more complex, rapid sequences the sensory equipment is too slow for direct feedback correction, and adaptation is based on means for selection and regeneration of successful patterns of behavior for use in subsequent situations, i.e., on an internal dynamic world model.

At a higher level of conscious planning, most human activity depends upon a rather complex sequence of activities, and feedback correction during the course of behavior from mismatch between goal and final outcome will therefore be too inefficient because in many cases it would lead to a strategy of trial and error. Human activity in a familiar environment will not be goal-controlled; rather, it will be oriented toward the goal and be controlled by a set of rules that have proven successful previously. In unfamiliar situations when proven rules are not available, behavior may be goal-controlled in the sense that different attempts are made to reach the goal, and a successful sequence is then selected. Typically, however, the attempts to reach the goal are not performed in reality, but internally as a problem solving exercise–i.e., the successful sequence is selected from experiments with an internal representation or model of the properties and behavior of the environment. The efficiency of humans in coping with complexity is largely due to the availability of a large repertoire of different mental representations of the environment from which rules to control behavior can be generated ad hoc. An analysis of the form of these mental models is important to the study of human interaction with complex man-made systems.

Basically, meaningful interaction with an environment depends upon the existence of a set of invariate constraints in the relationships among events in the environment and between human actions and their effects. The implications of the discussion above is that purposive human behavior must be based on an internal representation of these constraints. The constraints can be defined and represented in various different ways, which in turn can serve to characterize the different categories of human behavior.

Skills, Rules, and Knowledge

When we distinguish categories of human behavior according to basically different ways of representing the constraints in the behavior of a deterministic environment or "system," three typical levels of performance emerge: skill-, rule-, and knowledge-based performance. These levels and a simplified illustration of their interrelation are shown in Figure 9.1.

Skill-based behavior represents sensorimotor performance during acts or activities that, after a statement of an intention, take place without conscious control as smooth, automated, and highly integrated patterns of behavior. Only occasionally is performance based on simple feedback control, where motor output is a response to the observation of an error signal representing the difference between the actual state and the intended state in a time-space

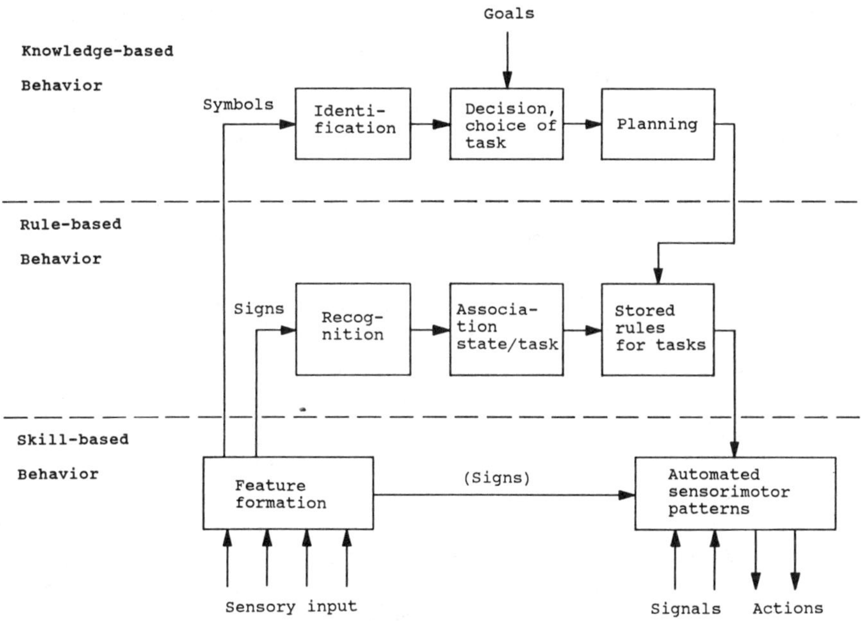

Figure 9.1. Simplified diagram of the three levels of control of human actions. The three levels in a real situation interact in a much more complex way than shown. Note that the arrows indicate information flow, not necessarily control. For instance, only the lowest level can be considered "data-driven." For the higher levels, information is actively sampled ("rule-" or "model-driven"). However, two aspects are necessary in explaining human behavior: reasons in terms of intentions, conditioning the organism top-down, and causes in terms of sensory information releasing actions bottom-up, as shown in the figure. For control of attention, see Chapter 8. [Reproduced from Rasmussen (1983) with permission from IEEE.]

environment, and where the control signal is derived at a specific point in time. Typical examples are experimental tracking tasks. In real life, this mode is rarely used, and only for slow, very accurate movements–assembly tasks, drawing. In most skilled sensorimotor tasks, the body acts as a multivariable, continuous control system synchronizing movements with the behavior of the environment. As discussed in the previous chapter, performance is then based on feed-forward control and depends upon a very flexible and efficient dynamic internal world model.

It is characteristic that skilled performance rolls along without conscious attention or control. The total performance is smooth and integrated, and sense input is not selected or observed–the senses are only directed toward the aspects of the environment needed to update and orient subconsciously the internal map. The man looks rather than sees.

In some cases, performance is one continuous, integrated dynamic whole, such as bicycle riding or musical performance. In these cases, the higher-level control may take the form of conscious intentions to update the dynamic world model and thereby to "modulate" the skill in general terms, such as "be careful now the road is slippery" or "watch out, now comes a difficult passage." In other cases, performance is a sequence of rather isolated, skilled routines that are sequences of a conscious "executive program." In general, human activities can be considered as a sequence of such skilled acts or activities composed for the actual occasion. The flexibility of skilled performance is due to the ability to compose from a large repertoire of automated subroutines the sets suited for specific purposes.

At the next level of *rule-based behavior*, the composition of such a sequence of subroutines in a familiar work situation is typically consciously controlled by a stored rule or procedure that may have been derived empirically during previous occasions, communicated from other persons' know-how as an instruction or cookbook recipe, or it may be prepared on occasion by conscious problem solving and planning. The point here is that performance is goal-oriented, but structured by "feed-forward control" through a stored rule. Very often, the goal is not even explicitly formulated, but is found implicitly in the situation releasing the stored rules. The control is teleologic in the sense that the rule or control is selected from previous successful experiences. The control evolves by the survival of the fittest rule. In effect, the rule will reflect the functional properties that constrain the behavior of the environment, but usually properties found empirically in the past. Furthermore, in actual life, the goal will only be reached after a long sequence of acts, and direct feedback correction considering the goal may not be possible. Feedback correction during performance will require functional understanding and analysis of the current response of the environment, which may be considered an independent, concurrent activity at the next higher level (knowledge-based).

The boundary between skill-based and rule-based performance is not quite distinct, and much depends on the level of training and on the attention of the person. In general, the skill-based performance rolls along without the person's conscious attention, and he will be unable to describe how he controls and on what information he bases the performance. The higher-level, rule-based coordination is in general based on explicit know-how, and the rules used can be reported by the person, although the cues releasing a rule may be difficult to describe.

During unfamiliar situations, faced with an environment for which no know-how or rules for control are available from previous encounters, the control of performance must move to a higher conceptual level, in which performance is goal-controlled, and *knowledge-based* (knowledge is here taken in a rather restricted sense as possession of a conceptual, structural model). The level might more appropriately be called model-based. In this situation, the goal is explicitly formulated, based on an analysis of the environment and the overall aims of the person. Then, a useful plan is developed–by selection, such that different plans are considered and their effect tested against the goal; physically by trial and error; or conceptually by means of understanding of the functional properties of the environment and prediction of the effects of the plan considered. At this level of functional reasoning, the internal structure of the system is explicitly represented by a "mental model" that may take several different forms. We will return to this point in the next chapter.

Similar distinctions between different categories of human behavior have previously been proposed. Fitts and Posner (1962) distinguish between three phases of learning a skill: the early or cognitive phase, the intermediate or associative phase, and the final or autonomous phase. If we consider that in real life a person will meet situations with a varying degree of training when performing his task depending on variations and disturbances, the correspondence with the three levels in the present context is clear.

These three levels of control of human behavior are closely related to the rational processing in the ladder of abstraction itself, in the habitual bypasses and the automated subroutines that form the elements of the task procedure. This relationship is illustrated in Figure 9.2.

Whitehead (1927), discussing symbolism, operates with three categories of human performance: instinctive action, reflex action, and symbolically conditioned action, which are also related to the present discussion:

> Pure instinct is the most primitive response which is yielded by organisms to the stimulus of their environment [p. 92].... Reflex action is a relapse towards a more complex type of instinct on the part of an organism which enjoys, or has enjoyed, symbolically conditioned action [p. 94].... Reflex action arises when, by the operation of symbolism, the organism has acquired the habit of action in response to immediate sense-perception, and has discarded the symbolic enhancement of causal efficacy [p. 96].... In symbolic conditioned action the causal efficacy is thereby perceived as analysed into components with the locations in space primarily belonging to the sense-perceptions [p. 94].... Finally mankind also uses a more artificial symbolism, obtained chiefly by concentrating on a certain selection of sense-perceptions, such as words for example. In this case there is a chain of derivations of symbol from symbol whereby finally the local relations between the final symbol and the ultimate meaning are entirely lost. Thus these derivative symbols, obtained as they were by arbitrary association, are really the result of reflex action suppressing the intermediate portions of the chain [p. 98].

Whitehead's discussion of symbols and derived symbols, the meaning of which is lost, leads to the distinction between signals, signs, and symbols.

Signals, Signs, and Symbols

One aspect of the categorization of human performance in skill-, rule-, and knowledge-based behavior is the role of the information observed from the environment, which is basically different in the different categories. The fact that

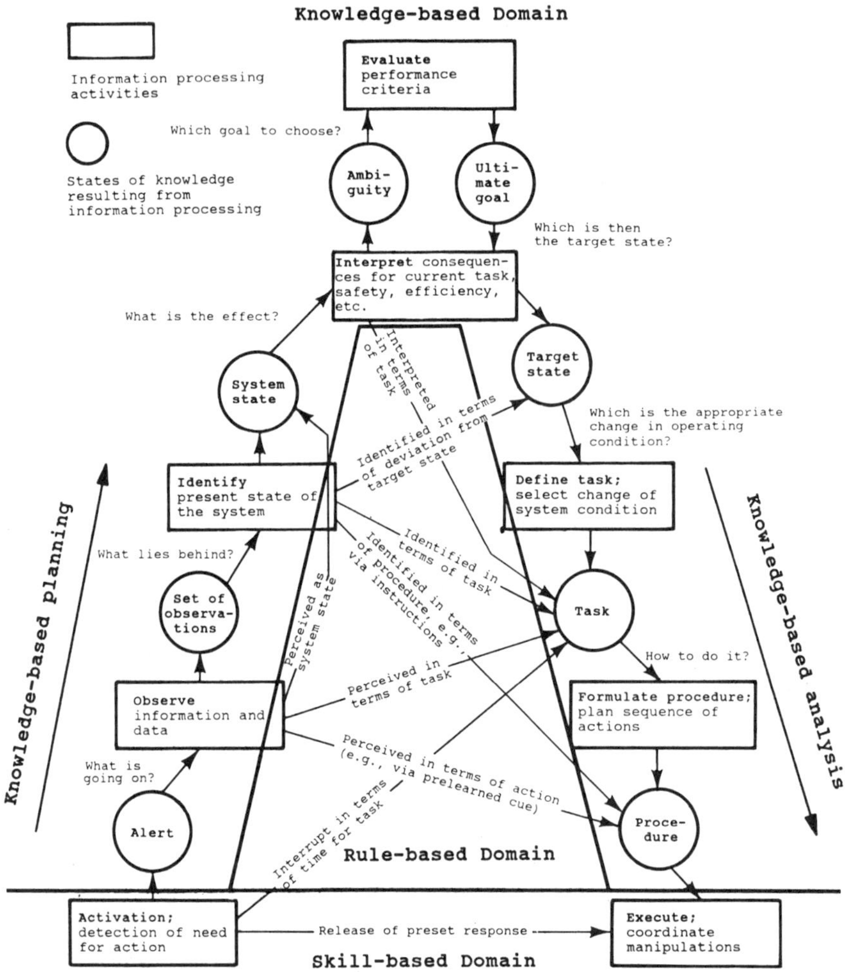

Figure 9.2. The relationship between the decision ladder and the three levels of control of human behavior.

information or indications from the environment can be perceived in basically different ways by a human observer is no new discovery (see, e.g., Morris, 1971; Eco, 1979), but curiously enough, it has so far not been explicitly considered by human-machine interface designers. This is the case even though major problems during unfamiliar situations may be caused by the fact that the same indication may be perceived in various different roles and that it is a well-known psychological phenomenon that shift between different modes of perception is difficult (functional fixation).

At the skill-based level, the perceptual motor system acts as a multivariable, continuous control system synchronizing the physical activity, such as navigating the body through the environment and manipulating external objects in a time-space domain. For this control, the sensed information is perceived as time-space *signals*, continuous, quantitative indicators of the time-space behavior of the environment. These signals have no "meaning" or significance except as direct physical time-space data. The performance at the skill-based level may be released or guided by value features attached by prior experience to certain patterns in the information not taking part directly in the time-space control but acting as cues or *signs* activating behavioral routines of the organism. At the skill-based level, the performance is similar to that of a data-driven continuous system, conditioned by signs in terms of patterns in the sensory information or intentions, or both, supplied from the higher levels of control.

At the rule-based level, the information is typically perceived as *signs*. The information perceived is defined as a sign when it serves to activate or modify predetermined actions or manipulations. Signs refer to situations or proper behavior by convention or prior experience; they do not refer to concepts or represent functional properties of the environment. Signs are generally labeled by names that may refer to states or situations in the environment or directly to goals and tasks of a person. Signs can only be used to select or modify the rules controlling the sequencing of skilled subroutines; they cannot be used for functional reasoning, to generate new rules, or to predict the response of an environment to unfamiliar disturbances.

To be useful for causal functional reasoning in order to predict or explain unfamiliar behavior of the environment, information must be perceived as *symbols*. Whereas signs refer to percepts and rules for action, symbols refer to concepts tied to functional properties and can be used for reasoning and computation by means of a suitable representation of such properties. Signs have external reference to states of and actions upon the environment, but symbols are defined by and refer to the internal, conceptual representation that is the basis for reasoning and planning. Cassirer notes (1944):

> Symbols—in the proper sense of the term—cannot be reduced to mere signs. Signs and symbols belong to two different universes of discourse: a sign is part of the physical world of being, a symbol is part of the human world of meaning.

The difference between signs and symbols, and the difficulty in the shift from rule-based reliance on signs to knowledge-based use of symbols is clearly illustrated in the testimony of the Three Mile Island operators to the U.S. Congress (Oversight Hearings, 1979):

> Mr. Frederick: Let me make a statement about the indications. All you can say about them is that they are designed to provide indications for whatever anticipated casualties you might have. If you go out of the bounds of an anticipated casualty, if you go beyond what the designers think might happen, then the indications are insufficient and they may lead you to make wrong inferences. In other words, what you are seeing on the gauge, like what

> I saw on the high pressurizer level, I thought it was due to excess inventory. In other words, I was interpreting the gauge based on the emergency procedure, where the emergency procedure is based on the design casualties. So the indications then are based upon my interpretation. Hardly any of the measurements that we have are direct indications of what is going on in the system. They are all implied measurements [p. 138].

If to this is added the difficulty in abandoning a search for a rule, which is not there, the point becomes clear.

> Mr. Faust: What maybe you should try to understand here is that we are trying to gain the proper procedure to go at it. We were into possibilities of several procedures, not just one, to cover what was happening. It has not been written, in fact. So we were still trying to determine which procedure to go by [p. 139].

The distinction between the perception of information as signals, signs, or symbols is generally not dependent on the form in which the information is presented, but rather on the context in which it is perceived, i.e., upon the intentions and expectations of the perceiver. Whorf expresses this well-known fact in the following way (Whorf, 1956):

> The categories and types that we isolate from the world of phenomena we do not find because they stare every observer in the face; on the contrary, the world is presented to us in a kaleidoscopic flux of impressions which has to be organized by our minds....

Figure 9.3 illustrates how the same instrument can serve to transmit all three kinds of message.

The discussion of the different perception of information is a classic topic within biology and philosophy, and similar distinctions have been drawn. Dewey and Bentley (Bentley, 1950) apply the same definition for sign and symbol as discussed above, but use the term "signal" in a different way, which is more related to its use in classic discussions of reflexive behavior such as that of Pavlov's dogs: they...

> have employed the word "sign" to name this technically characteristic "indirectness" as it is found across the entire behavioral field.... Within the range of sign, the word "signal" was chosen to name the underlying sensory-perceptive level; the word "designation" for the next higher evolutionary level–namely, that of linguistic sign operation; and the word "symboling" for a still higher range in the evolutionary sense....

In the present human-machine context, it seems to be important to keep the role of information as time-space signals, which are processed directly in a dynamic control of the motor performance, separate from the role as signs that serve to modify actions at a higher level.

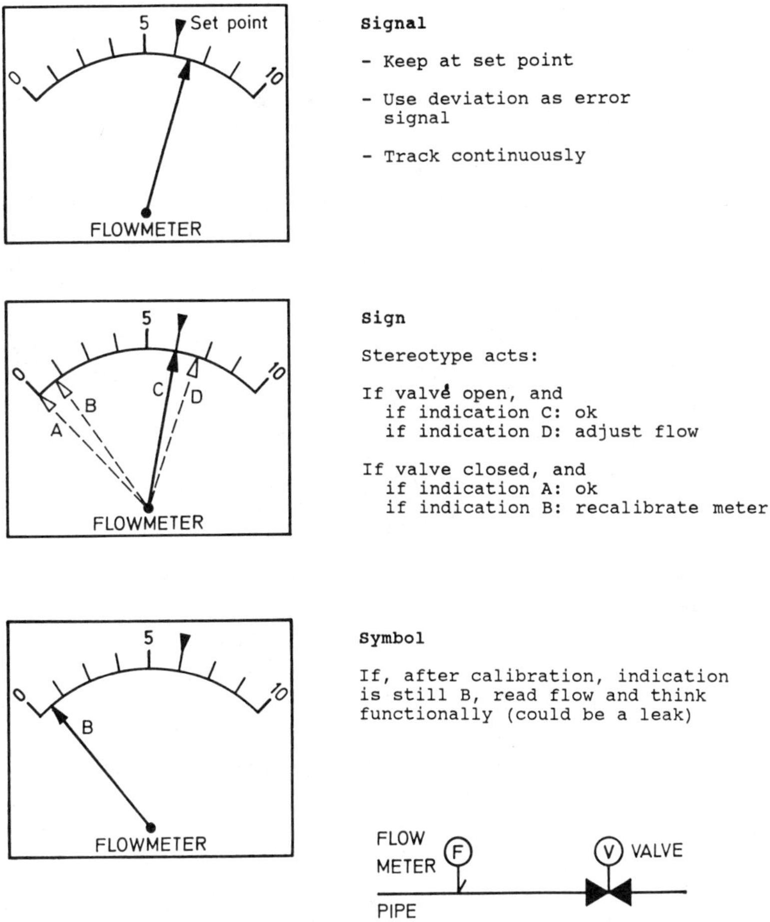

Figure 9.3. The same indications can be interpreted by a process operator as a signal, a sign, or a symbol depending on circumstances. [Reproduced from Rasmussen (1983) with permission from IEEE.]

The distinction signs/symbols is also treated by von Foerster (1966). He focuses, however, upon the difference between humans and animals:

> Communication among social insects is carried out through unalterable signs which are linked to the genetic make-up of the species. To communicate acquired knowledge by passing through generations, it must be communicated in symbols and not signs. This separates man from beasts.

Sometimes, but operating from signs may also be the normal way to be efficient for humans. You act on signs, but understand symbols.

To sum up, the three levels of behavior in the present context are characterized by different uses of the information available, and the distinction is very clear from an information processing point of view:

Signals are sensory data representing time-space variables from a dynamic, spatial configuration in the environment, and they can be processed by the organism as continuous variables.

Signs indicate a state in the environment with reference to certain conventions for acts. Signs are related to certain features in the environment and the connected conditions for action. Signs cannot be processed directly; they serve to activate stored patterns of behavior.

Symbols represent other information, variables, relations, and properties, and can be formally processed. Symbols are abstract constructs related to and defined by a formal structure of relations and processes, which by conventions can be related to features of the external world.

Semiotic Interpretation of Human Acts

In the discussion above, only the interpretation of the information picked up by humans has been discussed in semiotic terms. However, the same point of view can be taken on human actions. In human-machine interaction, the transfer of information from the human to the system will in general be in the form of actions on the system, i.e., it involves the change of position of a physical element of the system. In some cases this change of position is directly involved in the intended change in the environment, e.g., when assembling things, moving things, or shaping objects. In other cases, the movement itself is only indirectly involved since the system amplifies the movement or the applied force in order to have the intended change, e.g., when tools and manipulators are used, or the course of a vehicle is controlled. In both cases, the movements are acting as continuous signals in a time-space control loop of which the human body is an integrated part.

Frequently, however, the manual act or movement itself is not at all involved in the intended change in the environment, but merely acts as a sign that by convention and design is suited to release a predetermined causal chain in the environment as, for example, when a switch is activated in order to turn on the light or to start a car. In human-machine interaction, it therefore appears to be necessary to consider the same distinction between signals and signs for the significance of human acts as it is for the information observed by a human. Also, the interpretation of a movement or act as a symbol can be relevant, in particular if gestures are intended for interpretation by another human, not to mention speech acts. When acts or movements serve to transfer information to or control the processes of an automatic information-processing system, they may appear to be symbolic, but in general, they will only at the present state of development be signs releasing predetermined causal sequences and not participate in a symbolic process. Only for intelligent systems, which understand the meaning of messages (i.e., that perform under the control of conceptual models and goals, rather than rules), can input acts be interpreted as symbols. Compare this with Searle's (1981) discussion on whether computer programs are able to understand language.

The distinction is related to the phenomenologic question whether a human will consider tools and information gathering equipment to be extensions of the body or parts of the environment. This appears to be important for the resulting data-processing capacity of the total human-machine system. Different typical categories of human-machine interaction will therefore be considered in more detail. *Direct manipulation* of the physical environment takes place in manual tasks as, for example, when moving and shaping objects and assembling things. At the conscious level, objects are perceived directly in terms of their functional implications, and intentions are stated in terms of acts to perform, goals or targets to reach–generally not in terms of movements. Therefore, conscious statements of intention have the character of signs releasing acts, i.e., patterns of movements. The control of the necessary movement is based on the sensing of time-space signals, which aligns the internal model. In direct manipulation, both attention and intention are related to the time-space configuration of the environment but expressed at the level of acts, intentions, i.e., in terms of signs related to the state of environment by convention. They are typically not tied only to the specific situation, but applicable for many similar situations. However, the integrated interaction in the time-space control loop (the skilled patterns) is specific for the actual situation. Examples of general expression of intentions are "fill the bottle," "screw lid on," and "we have to put this together." The more general statements of intention are interfaced with the real physical environment by the sensorimotor skill, controlled by the internal world model; see Figure 9.4.

Figure 9.4. During highly skilled, direct object manipulation, the spatial-temporal continuous signal loop through the sensorimotor functions is intact. The loop is activated and the properties controlled by perception of cues as signs.

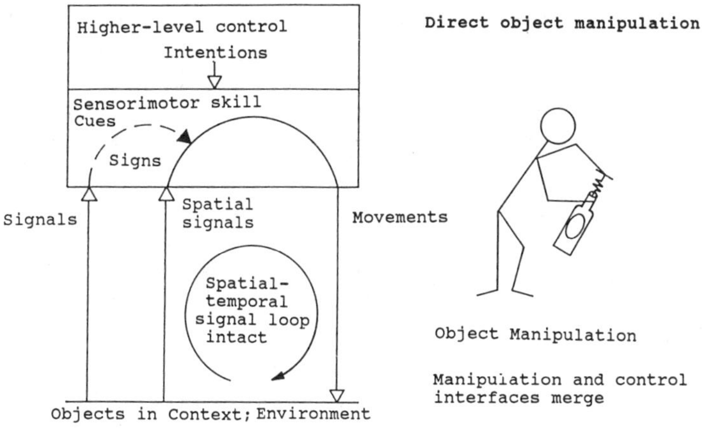

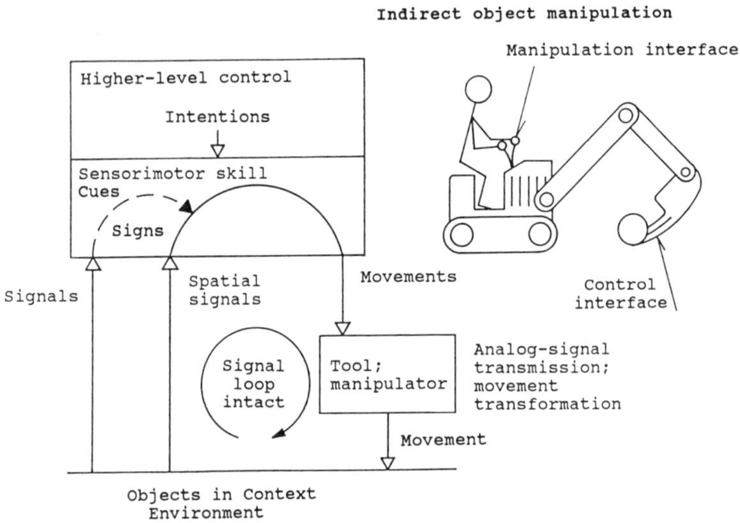

Figure 9.5. By indirect object manipulation, when the spatial-temporal signal loop is intact, the tool may be considered an extension of the motor system of the human. Manipulation and control interfaces are different.

Indirect manipulation appears when mechanical tools are applied to manipulate or transform objects. Examples can be hand tools and major machinery such as bulldozers or remote manipulators for radioactive hot-cells, the function of which can be monitored visually. The characteristic feature of this situation is that intention and attention are both focused on the task at the interface between the tool and the environment; the tools or manipulators are perceived as extensions of the human body because the time-space control loop through the sensorimotor system is intact. When the initial training period is over, the properties of the tool during movements, as, for example, dynamic properties, are integrated with those of the body into one whole pattern of the internal world model. This category also includes vehicle control tasks such as car driving and aircraft piloting, as long as they are based on direct visual perception. The prerequisite for this "embodiment" of the tool (Ihde, 1981) is the possibility of a dynamic integration of the control of body and tool movement, which implies that the tool is transmitting signals from the "manipulation interface" to the "control interface"; i.e., the focus of intention and attention is at the task itself, not at the tool operation. This also implies that the sensing channel must be able to transmit the time-space signals. The control signal loop is closed through the sensorimotor mechanisms, with a higher-level control performed by direct perception of functional properties and values and of signs that can be related to intentions (see Figure 9.5).

The introduction of means for information transmission between the task location and the human controller is necessary in many situations of remote

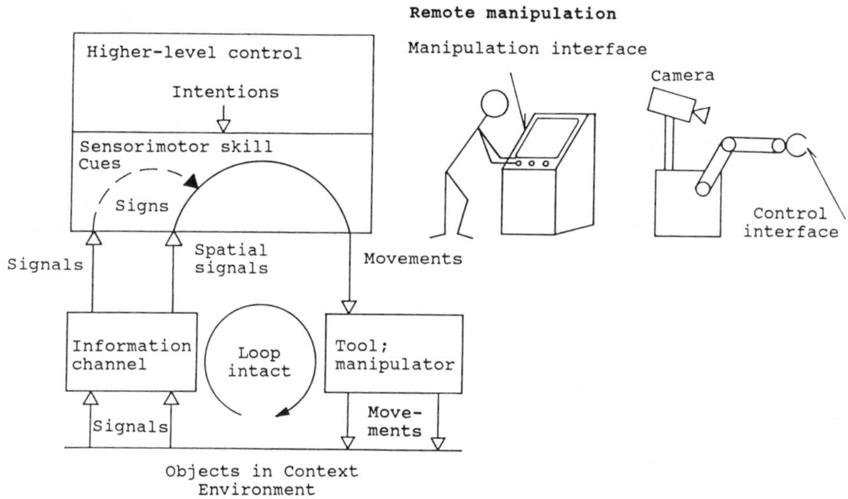

Figure 9.6. Remote manipulation. When an information channel is inserted for remote control, this channel may be considered an extension of the senses, if the spatial-temporal signal loop is maintained. Attention and control will be at the remote object interface. The skilled performance effectively takes part in the task itself.

manipulation, for example, for remote underwater manipulators, remote control of vehicles, and of robots for work in radioactive areas or for space work. When the information transmission channel is able to transmit the time-space signals necessary for control of movements as well as information carrying functional value features and signs for higher-level formation of intention–as, for example, will be the case for a closed circuit television channel or a microscope–the information channel will act as an extension of the sensory channel, and the attention and intentions can be focused on a control interface at the location of the remote task. The characteristics of this situation can be maintained even if the visual appearance of the work scene is not itself directly transmitted but only selected features, as long as the time-space signal information needed for movement control together with the features needed for interaction control are properly transmitted. Examples of such displays are the analog displays for instrument piloting of aircraft. In case of such abstract displays, the illusion of being on-site will generally disappear, and attention/intention level may shift to the display interface. The only way to distinguish whether attention is paid to the task itself or the representation by the display will be the language used to express intentions. With pictorial representation, distinction by means of the language terms disappears, and only subjective experience may indicate the focus of attention. In terms of control performance, the distinction is rather irrelevant because proper sensorimotor skill control can be maintained through the intact signal transmission loop in both cases; see Figure 9.6.

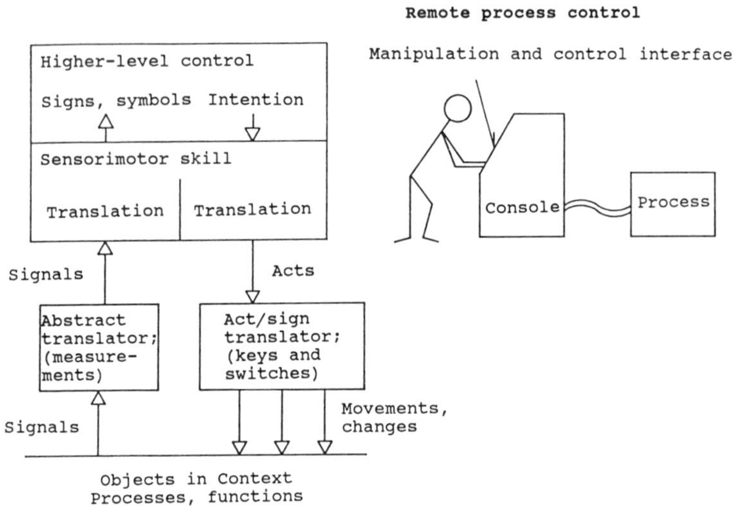

Figure 9.7. Remote process control. When symbolic indicators and discrete control keys changing parameters in an invisible process are inserted, the spatial-temporal loop is broken. Sensorimotor skill is no longer part of the primary task, but functions as a translation interface.

The time-space signal loop may be broken in some types of *remote process control*–the manipulation channel may not transmit space-time analog signals but, for example, alphanumerically coded orders, or the sensory channel, or both, may be coded in abstract codes rather than analog representation of the effect of acts. This is typically the situation if the task is not related to concrete space-time object manipulation, but to a higher-level state of a physical process that is not immediately visible, as, for example, the processes of a power plant or chemical process plant; see Figure 9.7. If the time-space control loop is broken in this way, there are translating functions in each transmission path, and the human attention/intention level will typically be at the surface of the system. During routine situations for operations based on signs and associations to acts (i.e., rule-based performance) upon the manipulation interface. During unfamiliar situations, the attention/intention must be time-shared between this interface (in order to translate readings to symbols and intentions to actions) and the mental representation (in order to make functional inferences in the knowledge-based domain). In this situation, the automated sensorimotor functions are applied to functions of translation that are unrelated to the higher-level cognitive task. The high-capacity sensorimotor skill will not be available to the main problem, but will be occupied by interface manipulation and sign recognition. Furthermore, during unfamiliar situations, cognitive resources must be spent on the translation tasks related to state identification and action planning.

The aim of introducing interface systems based on information technology should be to select a coding system that eliminates the human translation task and makes it possible to apply the high sensorimotor capacity directly in the main task itself. This implies at least two conditions. First, the presentation of information on the interface must be based on symbols related directly to the internal function to be controlled. Second, the symbols and structure must be chosen so as to allow operation by direct manipulation of the time-space configuration of the symbolic representation. In this case, there is a one-to-one analog mapping of the configuration of the symbolic representation and the function to be controlled, and the sensorimotor skill can be applied to the central task. This mean in fact that the time-space control loop is intact with a symbolic reinterpretation of the time-space signals in the translating human-machine interface, where the displayed graphic symbolic configurations are treated as artificial objects. This problem will be discussed in more detail in Chapter 10.

Skills, Rules, and Knowledge as Stages in Learning a Skill

In the three-level model, the final stage is the skill level where automated routines are based on subconscious time-space manipulations of objects or symbols in a familiar scenery. During training, the necessary sensorimotor patterns develop, while the activity is controlled by other means. From a control point of view, two problems must be solved during training. First, the activity must be synchronized with the behavior of the environment in order to meet the proper purpose. Second, the activity must be optimized to form a smooth and efficient pattern within the boundary of the task. A good example illustrating the synchronization and optimization problem is learning to ride a bicycle. The control of the activity during the initial phases of synchronization and formation of the sequence of acts can be obtained in various ways. It may happen directly at the skill level by imitation and trial-and-error, as, for instance, learning to play an instrument by ear or children learning to talk, walk, etc. In some cases, synchronization presents a problem–as in walking or riding a bicycle–and an external assistant with hand-on-the-saddle kind of support will be effective. In other cases, control at the rule-based behavioral level will be efficient during development of the automated skill. The rules may be obtained from an instructor or a textbook, as is typically the case in learning to drive a car, to operate tools and technical devices supplied with an instruction manual, or to manage social interactions from rules of good manners. And, finally, persons with a basic knowledge of the structure and functioning will be able to generate themselves a set of rules to control activities related to various purposes during early phases of learning.

An important point is that it is not the behavioral patterns of the higher levels that are becoming automated skills. Automated time-space behavioral patterns are developing while they are controlled and supervised by the higher-level activities, which will eventually deteriorate, and their basis as knowledge and rules may deteriorate. In fact, the period when this is happening may lead to errors due to interference between a not fully developed sensorimotor skill and a gradually deteriorated rule system. This kind of interference is known also to highly skilled musicians when they occasionally start to analyze their performance during fast passages. It seems to be plausible also that this effect can play a role for pilots of about 100 hours flying experience, which is known to be an error-prone period among pilots.

If rule-based performance is characteristic of professional activities in general, one would expect the basic causal or functional understanding to deteriorate. This is in fact the evidence found by Ackermann and Barbichon (1963) from their analysis of the organization of knowledge and the explanation of phenomena as presented by electrical and chemical technicians in industry. Based on analysis of interviews, their conclusions were that the professional knowledge of the technicians was fragmented, with lack of relationship among phenomena, with barriers between theory, practice, and extraprofessional life, and lack of relationship among various representations–mathematical, graphic, concrete, and analogical–of a particular phenomenon. One explanation could be, as the authors suggest, that theoretical knowledge is not used and the findings reveal rudimentary memories from basic education. But, seen in our context, it could also be that basic symbolic knowledge and representations are typically used to support stereotyped lines of functional reasoning in various typical situations, and causal reasoning therefore turns into rule- and skill-based manipulations of symbolic representations that will thereby lose their symbolic nature and theoretical relationship. As mentioned previously, this effect has been reported by Goldstein (1969) in solving algebraic equations. Analysis of the functional explanations offered by the technicians interviewed by Ackermann and Barbichon was characterized by resort to what the authors call verbal nominalisms, i.e., the use of technical terms without understanding their content or relationship, or verbal logicism, as, for instance, pseudoanalogies or replacement of logical sequences with chronologic sequences from the problem context. A general trend found is the replacement of functional arguments by reference to human interaction, a tendency that can be explained by the tight relationships between human acts and symbols that are degenerated into mere signs. Cuny (1971, 1973) studied the use of gestures by construction workers as, for instance, in control of crane movements, and describes the change of distinct, formally defined signs into more variate, analog signals, as skill develops.

During the first phases of skill acquisition, the activity will be controlled by separate cues that are defined individually and related to rules controlling very elementary acts. As skill develops, cues of more global nature in terms of the data pattern they include and depending on temporal and situational aspects will be adopted, and rules will be related to activity patterns rather than acts, i.e., intentions are expressed in terms of goals rather than acts to perform. Finally, the whole task is automated and is performed without conscious awareness as long as unexpected deviations do not occur, and the higher the skill, the less probable will unexpected occurrences be. In this phase, the initial expression of a rule as, "It's now 25 mph and I have to change gears," is replaced by the overall intention, "It's four o'clock and Friday, I have to go home," and the conscious mind can start planning the weekend activities while the skill brings you home through rush hours. This development of skill and the changing nature of the controlling rules have been studied by Dreyfus and Dreyfus (1980), who have formulated five distinct phases in professional learning which, however, appear to be rather ambiguous in their boundaries.

The example of driving home raises the question: how are the conscious reflections and the skilled performance interacting during performance with moderate and high skill? At low-skill levels, conscious attention will be occupied by the search for and analysis of cues and control of related actions. At high-skill levels, conscious attention may be occupied by completely different business. In situations of moderate difficulty, however, there will be time and capacity left for conscious attention from analysis of cues and situations and choice of rules, capacity useful for other activities. Knowledge-based analysis and monitoring of the rule-based control of activities may be important for coping with disturbances; see the discussion on human errors, Chapter 11. However, little is known about how the available time and conscious capacity are spent. Is conscious attention scanning through time to monitor the past and analyze the future? Sensorimotor skills depend on a properly updated internal world model; when a high degree of skill is developed, the integration across time is getting more and more extended. What is the immediate time span integrated in subconscious skill? What is the role in conscious evaluation of future developments for the updating of the internal model and its predictions?

Chapter 10

Mental Models: Aggregation, Abstraction, and Analogy

In the knowledge-based domain, the planning of interaction with the environment depends on structural knowledge, and on mental representations of the structural configuration of elements and their functional relationships. This is in contrast to the procedural knowledge that underlies rule-based behavior. As already discussed in relation to cognitive task analysis, these mental models can refer to properties at the various levels in an abstraction hierarchy. Like these different categories of models, different categories of strategies for operation with the models can be formulated–categories that are specifically tied to different types of models, tasks, goals, or subjective performance criteria; see the discussion of diagnostic strategies in Chapter 6. The efficiency of humans in coping with the complexity of the physical world is due to an ability to apply knowledge from previous experience to new situations by selecting and freely combining models, rules, and strategies that have proven successful separately in other situations. In the present section, the role of the different categories of mental models applied for knowledge-based operation will be discussed in some detail.

Several problems meet the human data processor in the interaction with a complex physical environment. Only a few elements of a problem can be within the span of conscious attention simultaneously. That is, the complex net of causal relations of the environment must be treated in a chain of mental operations, often leading to effects like the law of least resistance and the point of no return discussed in Chapter 5. That is, strategies that depend on sequences of simple operations are intuitively preferred, and there will be little tendency to pause in a line of reasoning to backtrack and develop alternative or parallel paths. An effective way to counteract limitations of processor capacity and short-term memory seems to be to modify the basis of mental data processing–the mental model–to fit it to the specific task in a way that optimizes the transfer of previous results and minimizes the need for new information. The efficiency of human cognitive processes seems to depend upon an extensive use of model transformations together with a simultaneous updating of the mental models in all categories with new input information, an updating that is performed below the level of conscious attention and control and depends on the context defined by the internal dynamic world model.

Several strategies for model transformation are possible and are generally used to facilitate mental data processing:

Aggregation: Elements of a representation are aggregated into larger units–chunks–within the same model category as familiarity with the context increases. This relates to the model transformation in the part-whole dimension; see Chapter 4.

Abstraction: The representation of the properties of a system or the environment in general is transferred to a model category at a higher level of abstraction; of particular importance here is transformations in the functional hierarchy along the means-end dimension; see Chapter 4.

Analogies–use of ready-made solutions: The representation is transferred to a category of the model for which a solution is already known or rules are available to generate the solution.

Aggregation

One way of coping with the complexity of the real-life environment is to change the resolution of one's considerations when the span of attention is increased, i.e., to aggregate physical items into higher-level objects or to integrate patterns of actions into routines controlled by higher-level intentions. Complexity is not an objective feature of a system (Rasmussen and Lind, 1981). The complexity perceived depends upon the resolution applied for the information search. A simple object becomes complex if observed through a microscope. Objective complexity can only be defined with reference to a given representation of a system. Therefore, the complexity faced by controllers is determined by the representation of the internal state of the system the interface allows the controller to develop for the various work conditions. Consequently, the apparent complexity of a system ultimately depends on the technology of the interface system. For instance, the complexity of the traditional industrial control consoles depends on the one-sensor-one-indicator technology. Only one level of resolution of the representation is available, and this has to meet the most detailed one needed in any situation. In that case, the interface must be complex by the law of requisite variety. One way to cope with control of systems that are complex in terms of large numbers of information sources and of devices for basic control actions is to structure the situation by aggregation, and thereby to transfer the problem to a level with less resolution.

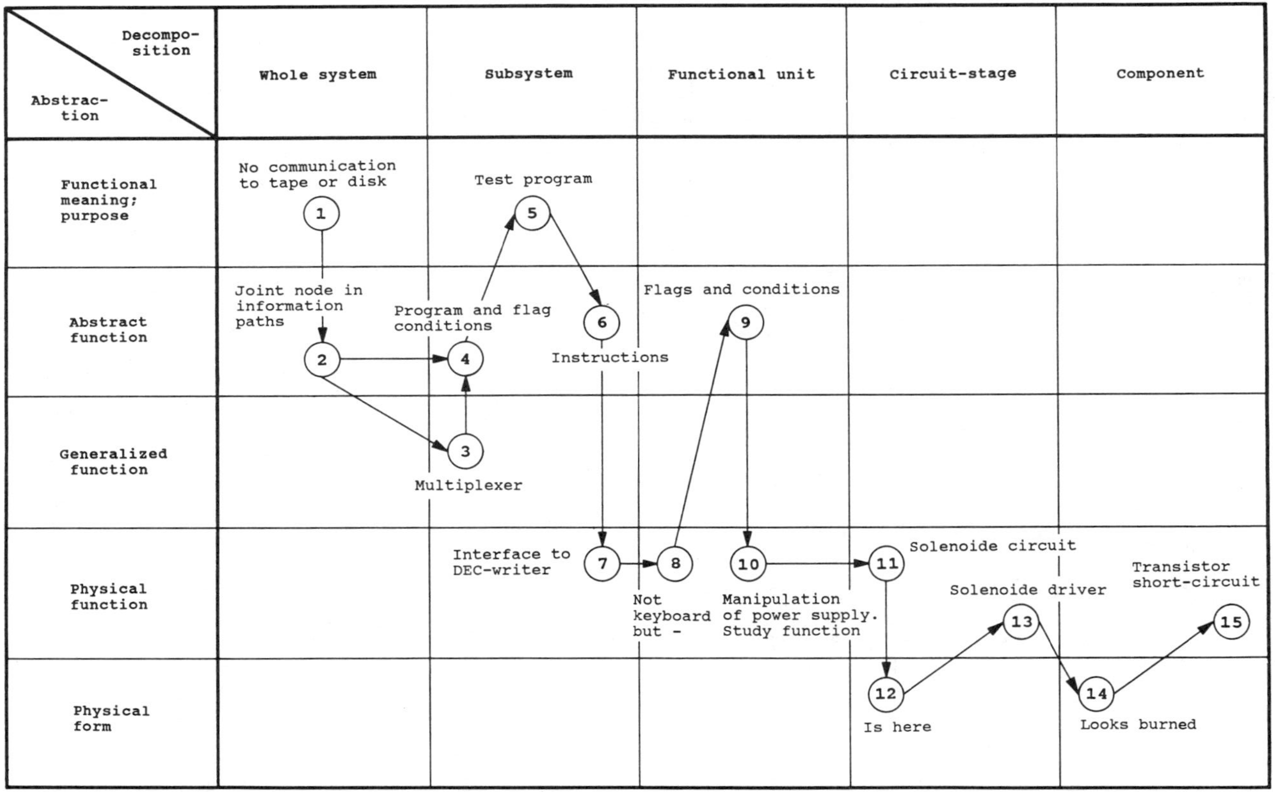

Figure 10.1. The focus of a computer troubleshooter's attention can vary with respect to the part-whole dimension and to the level of abstraction. The figure illustrates the trace followed in an actual diagnostic task.

Very often, a change of the level of physical decomposition within the span of attention is coupled with a change in the level of abstraction in representation but, basically, the whole-part and the abstract-concrete dimensions are conceptually separate. As an example, the changes in the two dimensions during troubleshooting in a computer system are shown in Figure 10.1.

Abstraction

It was discussed in Chapter 4 how a cognitive task can be formulated at several levels of abstraction in the representation of system properties. The same levels can therefore be expected in a description of the *content* of the mental models of human interaction with such systems. We will here only repeat some key features of the abstraction hierarchy, as a basis for discussion of its role in knowledge-based planning.

In the functional means-end *hierarchy*, the functional properties of the system are represented by concepts that belong to several levels of abstraction; see Figure 10.2 (which repeats Figure 4.1). The lowest level of abstraction represents only the physical form of the system–its material configuration. The

Figure 10.2. The functional properties of a system can be represented at several levels in an abstraction hierarchy.

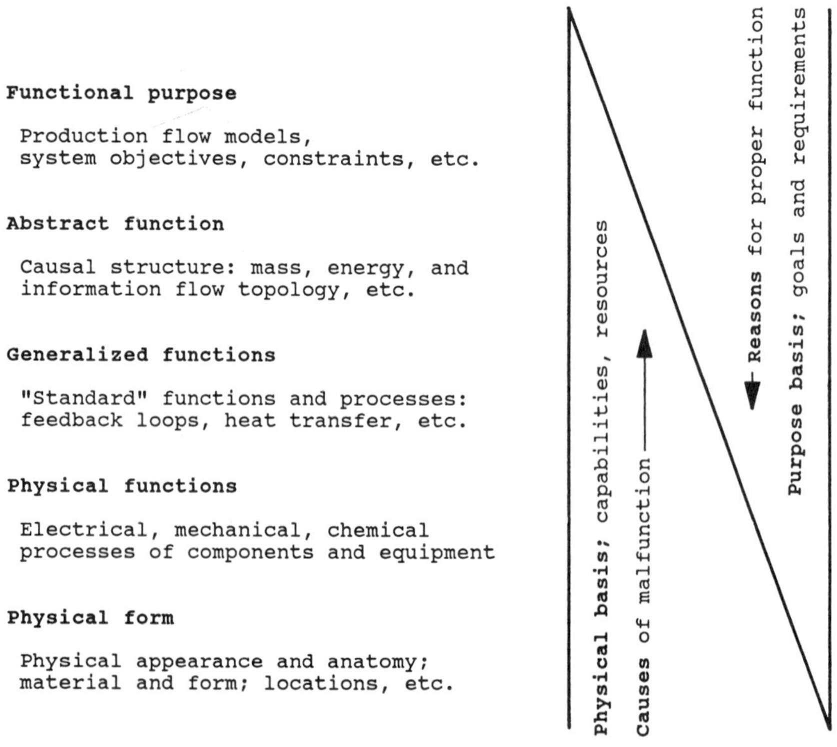

next higher level represents the physical processes or functions of the various components and systems in a language related to their specific electrical, chemical, or mechanical properties. Above this, the functional properties are represented in more general concepts without reference to the physical process or equipment by which the functions are implemented, and so forth. At the lower levels, elements in the process description match the component configuration of the physical implementation. When moving from one level of abstraction to the next higher level, the change in system properties represented is not merely removal of details of information on the physical or material properties. More fundamentally, information is added on higher-level principles governing the cofunction of the various functions or elements at the lower level. In man-made systems, these higher-level principles are naturally derived from the purpose of the system, i.e., from the *reasons* for the configurations at the level considered. A change of level of abstraction involves a shift in concepts and structure for representation as well as a change in the information suitable to characterize the state of the function or operation at the various levels of abstraction. Thus, an observer asks different questions of the environment depending on the nature of the currently active internal representation.

In other words, models at low levels of abstraction are related to a specific physical world that can serve several purposes. Models at higher levels of abstraction are closely related to a specific purpose that can be met by several physical arrangements. Therefore, shifts in the level of abstraction can be used to change the direction of paths, suitable for transfer of knowledge from previous cases and problems. At the two extreme levels of models, the directions of the paths available for transfer are in a way orthogonal, in that transfer at one level follows physical, material properties, while at the other it follows purpose.

As discussed in detail in Chapter 5, important human functions in human-machine systems are related to correction of the effects of disturbances and faults. Events can only be defined as disturbances or faults with reference to intended state, normal function, or other variants of system purpose or functional meaning. The functional models at the different levels of abstraction play different roles in coping with disturbed systems. Causes of improper functions are dependent upon changes in the physical or material world. Thus they are explained "bottom-up" in the levels of abstraction, whereas reasons for proper function are derived "top-down" from the functional purpose (see Figure 10.2). The clear difference between the propagation of causes of faults and reasons for function in the hierarchy has been discussed in detail by Polanyi (1967). This role of the abstraction hierarchy can clearly be seen in verbal protocols recorded during diagnostic search in information-processing systems. The diagnostician will frequently be forced to consider the functions of the system at several levels (see Chapter 4 for more detail). He will typically have to identify information flow paths and proper functional states by arguing top-down from the level of symbolic information, while he will utilize bottom-up considerations to analyze and explain the actual functional state from physical causes.

Another human task for which the use of representations at several levels of abstraction is of obvious value is the design of technical systems. Basically, system design is a process of iteration between considerations at the various levels rather than an orderly transformation from a description of purpose to a description in terms of physical form. There exists a many-to-many mapping between the levels; a purpose can be served by many physical configurations, and a physical system can serve many purposes or have a variety of effects. The use of different categories of representation in a design strategy has been explicitly discussed by Alexander (1964), pp. 89-90:

> Every form can be described in two ways: from the point of view of what it is, and from the point of view of what it does. What it is is sometimes called the formal description. What it does, when put in contact with other things, is sometimes called the functional description.... The solution of a design problem is really only another effort to find a unified description. The search for realization through constructive diagrams is an effort to understand the required form so fully that there is no longer a rift between its functional specification and the shape it takes.

If we accept the complex of strata between physical form and functional meaning of technical systems, an "invention" is related to a jump of insight that happens when one mental structure upward from physical form and another downward from functional meaning, which have previously been totally unconnected, suddenly merge to "a unified description."

Another consideration should be added to this discussion. Frequently, other persons will be part of the environment with which a particular person interacts, and for which he has to use mental models in order to cope with unfamiliar situations. As for technical systems, various levels of abstraction can be used to model human "functional" properties, and an analogy of the levels discussed in Figures 4.1 and 10.1 was drawn for "models of man" in Figure 4.3. All the levels are used in various research contexts, but what is of particular interest here is that, in ordinary working life, human interaction is based on a top-down prediction drawn from perceptions of other persons' intentions and motives, and on commonsense representations of human capabilities, together with knowledge of accepted practice. Causal bottom-up arguments literally play no role, and the most important information to use for planning human interactions for unfamiliar occasions is therefore knowledge of the value structures and myths of the work environment. The obvious reason for this is the complexity and flexibility of the human organism. However, it should be emphasized that, because of the growing complexity of information and control systems, the role of such "intentional models" (Dennett, 1971) is rapidly increasing, also for interaction with such technical systems. The distinction between causal and intentional systems with respect to problem solving will be discussed in more detail in the subsequent section on a means-end abstraction hierarchy.

Analogy; Transfer of Results and Rules

At each level of abstraction, reasoning depends on a particular type of model and rules for information processing. Therefore, shifting the level of modeling can be very effective in a problem situation because data processing at another level can be more convenient, the process rules can be simpler or better known, or results can be available from previous cases. A special instance of this strategy is the solution of a problem by analogy, which depends upon the condition that different physical systems have the same representation at higher levels of abstraction. Higher-level models for one physical configuration may therefore be reinterpreted to solve problems related to a quite different, unfamiliar configuration.

Use of different levels in an abstraction hierarchy has different implications for different kinds of problem environments, which will be considered in more detail in the next section.

In some cases, efficient strategies can be found where symbols are transferred to another level of abstraction and reinterpreted. A simple example will be the subconscious manipulation of symbols that are reinterpreted as artificial objects, e.g., Smith's (1976) solution of scheduling problems by manipulation of rectangles; or the reinterpretation of numbers in terms of actions for calculations by means of an abacus. This kind of problem-solving skill will be discussed separately later in the chapter.

Problem Solving in a Means-Ends Abstraction Hierarchy

In a task related to control of a physical system, control may be based on rules for action upon the system at a specific level of abstraction. Rules may be prescribed or found empirically at that level or developed by inference up or down the hierarchy. This may involve generation of new rules or transfer of known rules by analogy. Problem solving not related to a causal, physical system differs in this respect, and different categories of problem-solving environment can be identified.

Physical systems with known and invariate internal structure are responding to changes and to human acts according to basic laws of nature that therefore can be used to predict their behavior. They are here called *causal systems*, and their response to physical changes for which no experience is available for an observer can be explained or predicted by means of bottom-up reasoning in the abstraction hierarchy, i.e., by functional analysis.

This approach is not possible for all the environments in which humans have to make decisions. Systems with a high degree of autonomous internal functioning, with self-organizing and highly adaptive features, will change their internal functional organization continuously in order to meet the requirements of the environment and to suit their internal goals or performance criteria. Even

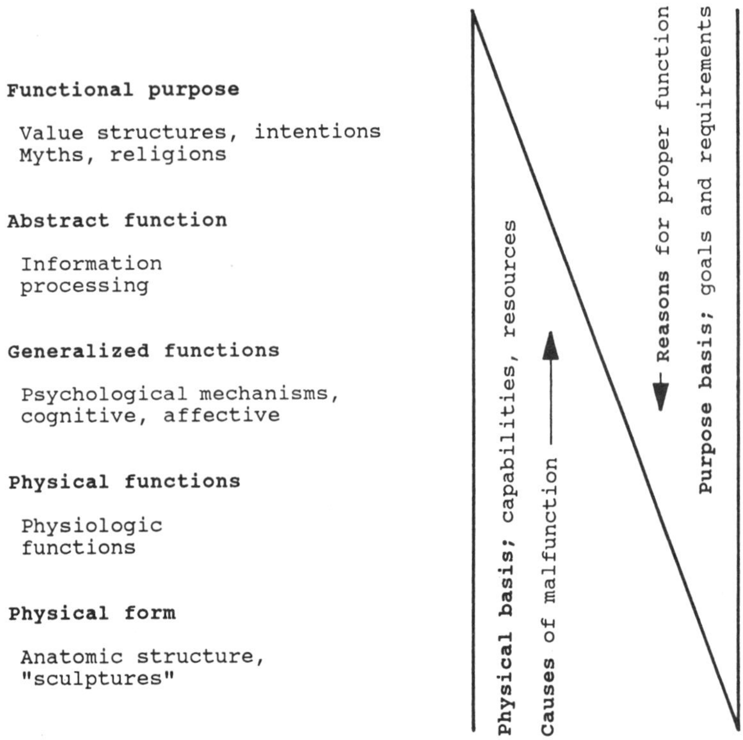

Figure 10.3. The abstraction hierarchy used for description of human functional properties.

though such systems are basically controlled by laws of nature, their complexity in general makes it impossible to explain or predict their performance by functional analysis during real-life decision making. The alternative is to consider such systems as *intentional systems* controlled by motives or intentions together with the constraints on performance posed by the environment–physically or in the form of conventions and legal requirements–and by the limiting capabilities of their internal mechanisms.

Prominent examples are humans and social systems. For humans, an abstraction hierarchy similar to that of Figure 10.3 is considered (which repeats Figure 4.3), with the different levels representing models in terms of anatomy, physiology, psychology, information processing, and value structures such as myths and religions. It has been a classic discussion since Stuart Mill whether the problem of relating mental events to physiologic states is due only to the complexity of the human organism and will be resolved by means of natural science in terms of causal explanations bottom-up in the hierarchy, or whether basically different, intentional models are needed in principle (Winch, 1958). It is, however, not only a problem in living organisms. Many technical systems, such as control systems and information processing systems, are very complex and have no simple relationship between their basic physical processes and their function in the information domain. Therefore, predictions regarding their behavior are more readily made when considering the systems as intentional systems (Dennett, 1971). Even in relatively simple systems, operators can be seen in verbal protocols to develop an explanation of system behavior from a top-down "redesign" of a reasonable functional structure from its supposed purpose, rather than to collect information on its actual, physical structure.

Decision making in control of intentional systems is based on knowledge of the value structures of the system, the actual input from the environment of the system, and its internal, limiting properties–i.e., it is based on reasoning top-down in the abstraction with little or no consideration of the internal causal structures or functions. This is probably the reason why top-level executive decision makers, according to Mintzberg's study (Mintzberg, 1973), do not behave according to analytical decision models, but prefer live action and constant consumer contacts instead of analysis of abstract reports, and current information–even gossip and hearsay–for statistics and status reports. Meeting people and considering hearsay are probably the best sources of information on current trends in value structures. This aspect is as important for understanding managerial decision making as the role of concrete, situational intuition of experts, which has been emphasized by Dreyfus (1980).

The strong emphasis on the causal relationships in the modeling of social systems, such as Forrester's world model (1971), rather than careful consideration of the dynamics of value structures, may greatly decrease the quality of long-term projections. A similar critique of traditional historical theories has been expressed by Toynbee (1972), who has a system-oriented approach to the history of societies. He defined a society in the following way: "A society is the total network of relations between human beings. The components of society are thus not human beings but relations between them" (p. 43). Presenting his "challenge-and-response model," he argues: "In my search up to the present point, I have been experimenting with the play of soulless forces–vis inertia and race and environment–and I have been thinking in the deterministic terms of cause-and-effect," and he continues: "The effect of a cause is inevitable, invariable, and predictable. But the initiative that is taken by one of the other of the live parties to an encounter is not a cause, it is a challenge. Its consequence is not an effect, it is a response. Challenge-and-response resembles cause-and-effect only in standing for a sequence of events" (p. 97).

The distinction between physical, causal systems and intentional, self-organizing systems must also be considered when research in human performance in games is used to explain performance faced in physical systems. In two-person games like chess, a person faces a system that is not controlled by basic, invariant laws, but by the intentions and value structures of the opponent. The game itself only represents a means of communication and the rules of the game serve only to constrain the decisions of the players to a well-defined set in each situation. Decisions depend upon prediction of the opponent's value structures and performance criteria and the strategy he adopts in the game.

The difference between games like chess and other social system contexts for management decision making is largely a question of formal consistency and invariance of the rule set. In games, the set of rules at the problem level is small and closed, and only the strategies for generation of proper action sequences are flexible and depend on top-down inferences regarding the opponent's intentions. In management decision making, there is room for invention of new rules of the game within the constraints of laws and "fair play."

Formal systems for decision making are problems that are only defined at one single level of abstraction, such as geometrical theorem proving and construction, cryptographic problems, puzzles, and purely logic problems; see, for instance, Newell and Simon (1972). The problem is stated here as an initial state and a target state, and the task is to identify a sequence of allowed, formal transformations that will close the gap. In this category are also several of the "context-free" tasks that are used for problem solving and human-machine interface experiments. It should, however, be realized that problem-solving behavior may be very different in one-level formal systems and in a problem context of an abstraction hierarchy. This leads to a discussion of problem-solving strategies with reference to the abstraction hierarchy.

Problem-Solving Strategies

An illustrative example of the role of the abstraction hierarchy can be found when comparing a decision task that has to be performed in a one-level formal description with the performance when the context is also available. The difference may partly be due to the use of shifts in level of abstraction to find paths for transfer of solutions and strategies by analogy, but also due to support of memory and search for rules in terms of structures at other levels of abstraction. A good empirical piece of evidence is the experiment made by D'Andrade (discussed by Rumelhart and Norman, 1981) who repeats the experiment of Wason and Johnson-Laird (1972). This experiment was based on a set of cards representing a concept like: if one side shows a vowel, then the back side displays an odd number. A subject was given a sample of four cards and asked which to turn in order to test the hypothesis. The experiment was repeated with the same concept disguised in a bill-signing context: if the amount of a bill exceeds $50, the supervisor must sign the back side. The ratio of correct solutions in the two experiments was 13% to 70%. Rumelhart and Norman conclude:

> What is the difference here? Why do people appear not to understand the meaning of "if" in the first case and understand it nearly perfectly in the second? This is exactly the kind of effect expected if our knowledge is embedded in a relatively inaccessible procedural format rather than as general rules of inference. The first case of the label factory represents a relatively

> unfamiliar case in which we cannot rely on specific knowledge and must, therefore, rely on general reasoning processes. The second case more nearly approximates our "real life" problem solving situations. Once we can "understand" the situation, the conceptual constraints of our specific knowledge can be brought into play, and the problem readily solved. It is as if our knowledge representation already contains all the reasoning mechanisms ordinarily required.
>
> Thus, it would appear that the context dependencies inherent in the more procedural representational systems are also present in the human reasoning system.

This is a rather implicit way of explaining the difference. A more explicit explanation seems to be possible, considering the use of an abstraction hierarchy in the problem solving. In the first experiment, the problem solving is based on formal, logical arguments at only one level of abstraction, on syllogistic logic that requires manipulation of abstract symbols and storage of intermediate results in short-term memory. In the second experiment, the context defines an intentional system in which the effects of the different decisions can very easily be inferred at the higher levels. The reasons for proper states can be inferred top-down. The problem is solved by top-down model modification, by transferring to a model of "reasonable states of affairs." Rumelhart and Norman refer to "understanding," which in our context can be viewed as the ability to transfer the problem–upward toward a reason, or downward toward a cause–to a level where immediate intuition from experience is available.

The role of the abstraction hierarchy not only seems to be the transformation of a problem to a level at which a solution is more readily available, e.g., by use of analogical reasoning. The transformation between levels seems to be a powerful tool for functional reasoning. Formal logical reasoning appears to be "horizontal" reasoning within a single level of the hierarchy, based on the classic syllogistic reasoning about membership of exclusive categories. On the other hand, practical, functional reasoning is related to "vertical" transformations in the abstraction hierarchy.

The difference between classic, formal logic and practical functional reasoning has long been a topic for discussion within philosophy. And, mentioned in passing, the well structured nature of the abstraction hierarchy related to man-made physical systems might make them a better suited vehicle for resolving such questions than the all-encompassing "real world" normally considered by philosophers. Bosanquet (1920) criticizes the classic syllogistic model of reasoning and, typically, as examples for his arguments selects electrical circuits with fuses and Harvey's discovery of the function of the circulation of blood. Bosanquet discusses at length the difference between syllogistic reasoning, which is linear, and functional inference, which he calls "systemic." Bosanquet (1888, 1920) argues that the explanation depends on the relational network of a whole, and that linear syllogistic arguments therefore are inadequate. Similar arguments have recently been presented by Harman (1982). He distinguishes inference or reasoning, on the one hand, from proof or argument on the other. To him, reasoning is a process of trying to improve one's overall view and is a holistic process that puts the problem into context, whereas rules of argument or proof are local rules of logical implication. He relates the distinction to a syntactic or grammatical basis, to a distinction between elements of logical form and nonlogical content, a distinction that is similar to the present discussion of arguments within a model at a single level of abstraction, and reasoning across levels by transformation and modification of models; cf. Harman's (1976) comment–practical reasoning is concerned with what to intend; formal reasoning, with what to believe. Formal logic arguments are a priori true or false with reference to an explicitly defined model; functional reasoning is dealing with relationships between models, and truth depends on correspondence with the state of affairs in the real world.

Certain schools of psychology also emphasize the significance of the structure of a problem as a guide to thinking. From analysis of a number of problem-solving scenarios, Wertheimer (1945) argues that problem solving is a productive process that depends on the structure of the problem situation in a way that was neither considered in the then-prevailing associative theories, nor in classic logic. The process is not characteristic merely by piecemeal linking together elements by associations; the flow is controlled by whole-characteristics, a "sensible expectation about structural truth." He argues the need for a theory that goes directly to the structural nature of the process and states "the gist of the thesis: structural reasons become the causes of the process," and he goes on to discuss the difference between reasons and causes in control of the process.

Also, Selz (1922; for a good English review, see De Groot, 1965) argues against an association model of thinking. His work is interesting in the present context in that his arguments are based on an analysis of errors in thinking. His premise is that given the mechanism is associative, errors should include a large fraction of associations with no functional meaning or without connection with the task. This is not what he found. On the contrary, errors typically appeared to be results of solution trials with regard to the task, which is somewhat misconceived. Selz is very modern in his conception of problem solving procedures, which are determined by: (a) the intellectual personality, i.e., the repertoire of solving operation dispositions; (b) the features of the problem; and (c) the subject's intention. The course of thought process is controlled by the subject's "schematic anticipation" with the features of gap and tension, and four basic operations are involved, such as likeness evocation, abstraction, combination, and complex completion, which can all be readily related to operations in an abstraction hierarchy. Selz's distinction between "productive" and "reproductive" thinking is related to the distinction between knowledge-based and rule-based control of behavior. In productive thinking, he distinguishes between "finding the means" and "applying the means." Finding new means may involve: (a) reproductive abstraction of means that identifies means by top-down search in the hierarchy; compare Duncker's "by means of which" relation, mentioned below; (b) coincidental identification, which looks like Duncker's "suggestion from below"; and (c) identification of means from structural insight into the nature of the task, i.e., restructuring through understanding. In all, the work of Selz points to the importance of a well-structured representation of the problem context. Through De Groot's (1965) study of chess strategies, the work of Selz has influenced modern AI research, but mainly through work on games and formal problems (Newell and Simon, 1972).

The role of a multilevel abstraction hierarchy in problem solving is most explicitly seen in Duncker's (1943) research on practical problem solving related to physical, causal systems (radioactive tumor treatment and temperature-compensated pendula). Based on verbal protocols, Duncker describes how subjects go from the problem to a solution by a sequence of considerations where the items proposed can be characterized by a "functional value" feature pointing upward to the problem, and a "by means of which" feature pointing downward to the implementation of a solution; see Figure 10.4. The relation to the abstraction hierarchy as shown in Figure 10.2 is clear. He states: "The final form of a solution is typically attained by way of mediating phases, of which each one in retrospect possesses the character of a solution, in prospect that of a problem.... When one closely examines what Maier calls 'direction' (in thinking), it becomes clear that direction is nothing but the earliest phase of the solution, i.e., the reformulation of the problem as it initiates the solution process concerned." Therefore, direction in thinking is given by the structure of the available abstraction hierarchy with its gaps; compare this with the structural anticipation of Selz.

Yet another observation on the role of an abstraction hierarchy on understanding a mechanical device has been reported by Rubin (1920), who reports an analysis of his own efforts to understand the function of a mechanical shutter of a photographic camera. He finds that consideration of purpose or reason plays a major role in the course of arguments: he conceived all the elements of the shutter in the light of their function in the whole. He did not perceive the task to explain how the individual parts worked, but rather what their functions were in the whole. How they worked was immediately clear when their function was known. He mentions that he finds it an analytical task to identify the function of parts, the direction of thought being from overall purpose to the individual function (top-down considerations). The hypothesis necessary to control the direction is then readily available. This approach was found to have additional advantages: solutions of subproblems immediately have their place in the whole picture, and it is immediately possible to judge whether a solution is correct or not. In contrast, arguing from the parts to the "way they work" is much more difficult as a result of being a synthesis. Solutions of subproblems must be remembered in isolation and their correctness is not immediately apparent.

For comparision, readers with philosophical inclinations should read Kant's discussion of the problem faced by "dissectors of plants and animals":

> In fact, they can as little free themselves from this teleological proposition as from the universal physical proposition; for as without the latter we should

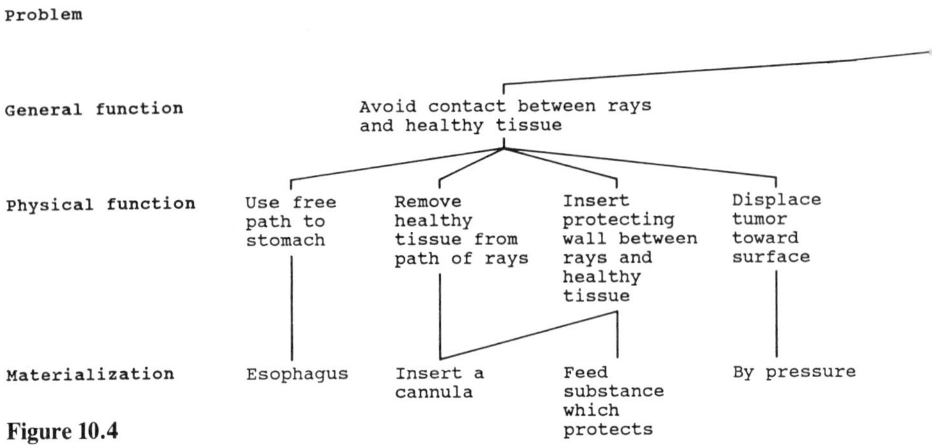

Figure 10.4

have no experience at all, so without the former we should have no guiding thread for the observation of a species of natural things which we have thought teleologically under the concept of natural purposes. [From Greene, Critique of Judgement, 1929, p. 471.]

Implications for the Design of Decision Support Systems

During new and unfamiliar system disturbances the task of a supervisory controller is very similar to the problem-solving task studied by, for instance, Duncker; the problem is formulated during the diagnostic phase and a solution is sought in terms of a reconfiguration of the available physical resources. The preceding discussion of the importance of the structure of the problem in terms of a means-end hierarchy therefore has a number of implications for analysis and design of systems for support of supervisory systems control.

Structuring Representation in Databases

In the design of human-machine interfaces for supervisory control as, e.g., industrial process control consoles, the emphasis has traditionally been on the presentation of measured data representing the physical state of the system and its processes. In complex systems, this information has been supplemented by information about the underlying functional structure by graphical means such as mimic diagrams, etc. This information is intended to serve the controller's identification of the actual state of processes bottom-up through the hierarchy. Information representing the intentions behind the system design, in terms of the purpose of functions and equipment and of constraints upon the acceptable

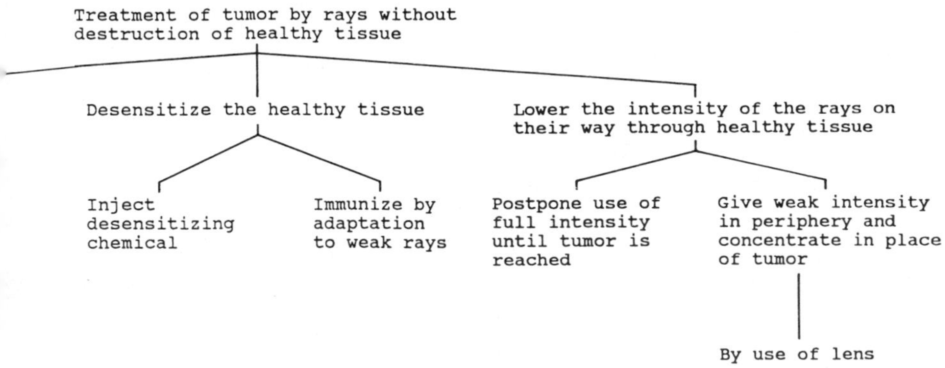

Figure 10.4. The generic solution tree found by Duncker in a problem-solving task. The levels are similar to the abstraction levels of Figure 4.1. [Adapted from Duncker (1945).]

operation, for instance, as derived from safety considerations, has only been very sparsely represented. This means that the interface system gives little or no support in the top-down derivation of the proper or acceptable states of processes. This kind of information was supposed to be immediately available to operators from their basic training. However, as systems become complex and potentially risky, and very low probabilities of erroneous decisions during rare events are required, this can no longer be assumed. During such situations, information about the purpose of interlocks, properties of equipment if used for untraditional purposes, etc., may be vital for ad hoc improvisations. Consequently, it becomes increasingly important to include in the support for decision making the information needed for top-down consideration of reasons. This means that the properties of the system to be controlled should be represented in terms of a consistent means-end hierarchy, systematically mapping the purpose/function/equipment relationships.

This leads to two problems: to find the information needed and to structure the representation. Information on "reasons" behind design decisions is to a great extent implied in standard practice, or only their implications are recorded by the designer in specifications and drawings. Regeneration may require quite a fair amount of work. For the structuring of the description, a systematic method based on a formal language is required. An approach to this problem has been taken by Lind (1982), who has developed a multilevel description of process systems in terms of their mass and energy flow topology (see Figure 10.5). This description is basically a representation of the flow topology of the system at several levels of decomposition, but it maps very well onto a means-ends hierarchy for a given system and is well suited for structuring a data base for supervisory support systems independently of the formats chosen for information presentation. In particular, this formal representation supports systematic

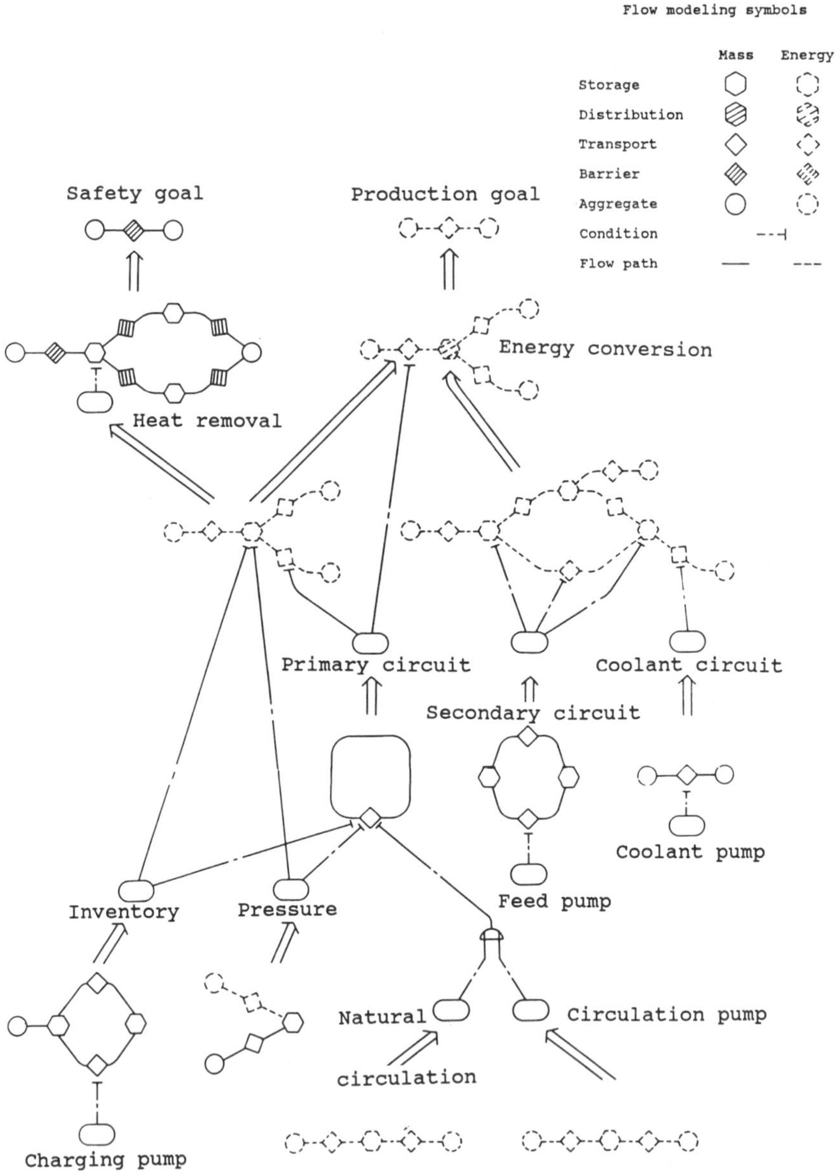

Figure 10.5. Illustrates the many-to-many mapping between the levels of purposes/functions/equipment of a nuclear power plant and the use by Lind (1982) of mass-energy flow topology for a systematic representation of the relationships.

analysis for the information gathering from the designers and, in addition, it will serve as a basis for automatic inference generation in the process computer's processing of the plant information.

Structuring Information Presentation

Given that the information is present in the database in a structured way, the problem is how to present it. During the various phases of a supervisory control task, the relevant level of consideration will vary in the means-ends hierarchy. An obvious proposal will be that the data from the system should be available in preprocessed form matching the level considered. For this purpose, the formal flow-term representation is well suited for data integration by the computer. In addition, the display format should match a useful mental model of the structure of the functions at that level. In this way, the information processing required by a supervisory controller for preparation of data to match his decision task can be greatly reduced. However, there is a price, which is the added information retrieval task of finding the relevant display in a perhaps large library. An experimental program is needed in order to analyze problem-solving behavior and subjective preferences in a task environment where information is available at several levels of a means-ends hierarchy and in several formats of presentation. Goodstein et al. (1983) have developed an experimental set-up based on a simulated nuclear power plant for this purpose. For these experiments, a set of displays based on the multilevel flow concept has been designed by Goodstein (1984) (see Figure 10.6).

The introduction of advanced information technology for support of supervisory control also leads to a cooperative decision making where parts of the underlying information processing are performed by computers. When it comes to more complicated solutions involving, for instance, automatic diagnosis or planning, a general problem is to reach compatibility between computer processing and human strategies to a degree that the decision support is actually accepted by operators. To reach such compatibility, computer processing should not be constrained to algorithmic, problem-specific processing, or heuristic processing based on procedural representation related to empirical evidence like that in the present typical expert systems. When empirical rules break down, the computer should be able to follow a human operator to higher-level, structural knowledge-based (model-based) processing.

In order to be able to add this level of knowledge-based control to expert systems (Barnett, 1982) and other complex decision support systems, it is necessary to formalize the description of the system to be controlled in an abstraction hierarchy, and to establish a set of rules for the vertical model transformations to be implemented in the computer program.

Predicting Decision-Error Modes

Another important implication of the functions of the means-ends hierarchy in control of problem-solving behavior is its influence on error modes during

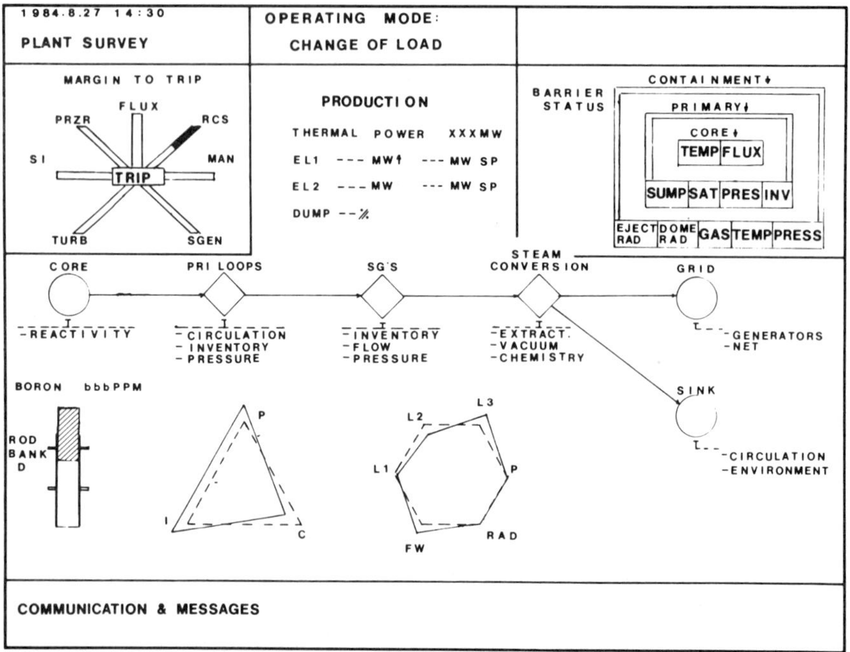
1984.8.27 14:30
PLANT SURVEY
OPERATING MODE:
CHANGE OF LOAD
MARGIN TO TRIP
FLUX
PRZR
RCS
SI
MAN
TRIP
TURB
SGEN
PRODUCTION
THERMAL POWER XXXMW
EL1 --- MW --- MW SP
EL2 --- MW --- MW SP
DUMP --%
BARRIER STATUS
CONTAINMENT
PRIMARY
CORE
TEMP FLUX
SUMP SAT PRES INV
EJECT RAD
DOME RAD
GAS TEMP PRESS
CORE
PRI LOOPS
SG'S
STEAM CONVERSION
GRID
-REACTIVITY
-CIRCULATION
-INVENTORY
-PRESSURE
-INVENTORY
-FLOW
-PRESSURE
-EXTRACT.
-VACUUM
-CHEMISTRY
-GENERATORS
-NET
SINK
-CIRCULATION
-ENVIRONMENT
BORON bbbPPM
ROD BANK D
P
I
C
L1
L2
L3
P
RAD
FW
COMMUNICATION & MESSAGES

(a)

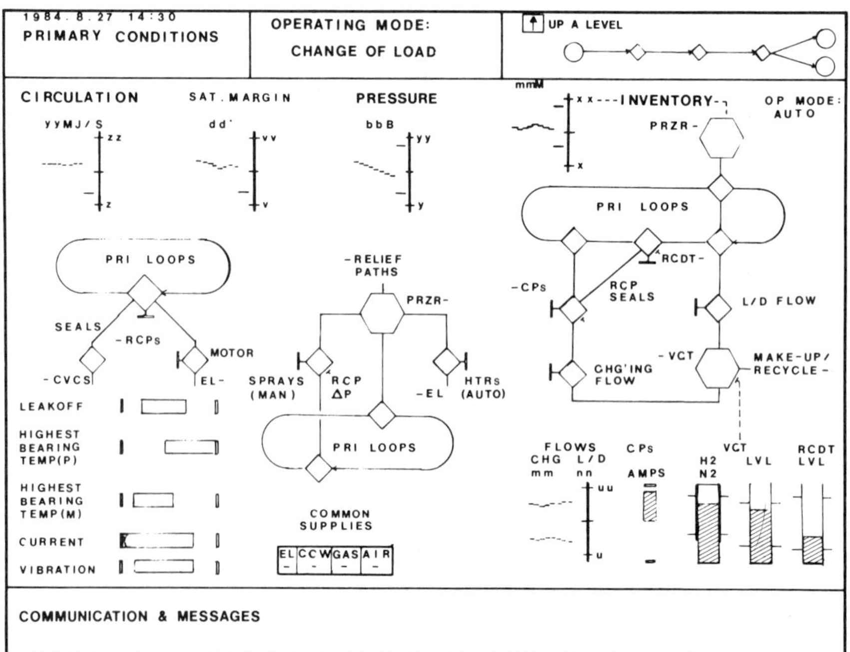
1984.8.27 14:30
PRIMARY CONDITIONS
OPERATING MODE:
CHANGE OF LOAD
UP A LEVEL
CIRCULATION
SAT. MARGIN
PRESSURE
INVENTORY
OP MODE: AUTO
PRZR-
PRI LOOPS
RCDT-
-CPs
RCP SEALS
L/D FLOW
CHG'ING FLOW
-VCT
MAKE-UP/ RECYCLE-
SEALS
-RCPs
MOTOR
-CVCS
EL-
-RELIEF PATHS
PRZR-
SPRAYS (MAN)
RCP ΔP
-EL
HTRs (AUTO)
LEAKOFF
HIGHEST BEARING TEMP(P)
HIGHEST BEARING TEMP(M)
CURRENT
VIBRATION
COMMON SUPPLIES
EL CCW GAS AIR
FLOWS
CHG L/D
CPs AMPS
VCT
H2 N2
LVL
RCDT LVL
COMMUNICATION & MESSAGES

(b)

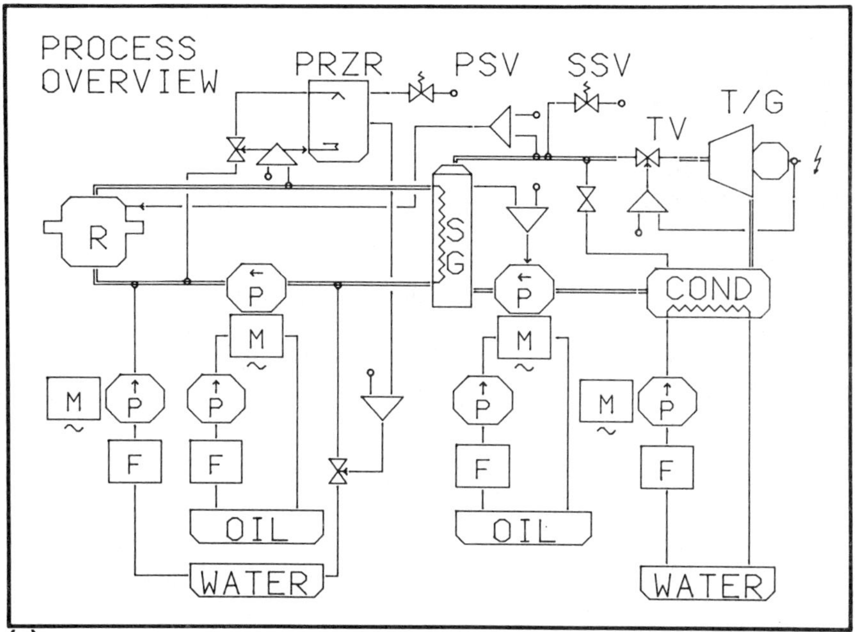

(c)

Figure 10.6. Computer-generated displays can be matched with the content and form of useful mental models at the different levels of abstraction. This figure illustrates display formats for control of a simulated nuclear plant, designed by Goodstein (1984) for experimental studies. One display (a) represents the overall energy flow topology of the plant, another (b) represents lower-level functions, while (c) represents the basic physical anatomy as a format for displaying sets of equipment status data.

decision making. As Selz noticed, decision errors are not stochastic events but depend upon the structure of the problem space in question. If supervisory decision making is considered as resource management in a task performed in a purpose/function/equipment hierarchy, several kinds of interference can be suggested as sources of systematically appearing decision errors. Equipment may be useful for different purposes in different situations or during different tasks, and conflicts may appear between the efforts of different users or the aims of the same person at different times. Seemingly unexplicable human acts during a critical task may be caused by mistakes caused by similarities of features at one or another level in representation. Similarly, decisions may be judged erroneous in retrospect, but in the actual situation be caused by attempts to test a very reasonable, but wrong, hypothesis. The conclusion of this is that a systematic representation of the means-ends relationships of the control object of a supervisory decision maker is a necessary prerequisite for modeling and prediction of decision errors.

Planning of Experiments

The distinction between problem solving, which is performed by formal logic within a problem space of only one level of abstraction, and problem solving based on transformation through the levels of an abstraction hierarchy has immediate implications for laboratory experiments on decision making. Frequently, more or less context-free problem representations are used for experiments on problem-solving strategies. These experiments are very effective when the aim is to model logical reasoning in a closed, formal system, but the implications for real-life tasks may be difficult to establish, because the setting does not invite a shift in the level of abstraction and hence does not support problem solving from prior experience based on functional inference and analogies. Furthermore, the lack of context leads to the replacement of the task to infer proper states from high-level purposes and the effect of complex disturbances from physical relationships, by the task to use the formal arguments of the one-level rules of the game. However, if embedded in various context scenarios, a context-free system simulation can be a very flexible experimental tool. Adding context to an experiment may change the task in two ways. It may add to the task the transformation of high-level criteria and physical disturbances into formal parameters, and it may change the decision strategy in the D'Andrade case. This means that, in addition to the traditional experimental psychology methodology, there is a need for experiments in much more complex and yet controlled settings (see, e.g., Goodstein et al., 1983).

Skills, Rules, and Knowledge in Problem Solving

In discussing knowledge-based problem solving in Chapter 9, it was briefly mentioned that problem solving can be transferred from the knowledge-based to the rule- or skill-based levels of behavior if it is possible to reinterpret the symbolic representations at another level of abstraction at which rules for manipulation are available. The potential of modern information and display technology for making graphic displays that act as externalizations of the symbolic mental representations suitable for human problem solving, in a form that can be directly manipulated, makes a more detailed consideration of this reinterpretation of symbols necessary.

The distinction between knowledge-based and rule-based behavior was also drawn by Duncker (1945) in his definition of "thinking": "Whenever one cannot go from a given situation to the desired situation simply by action, then there has to be recourse to thinking (by action we here understand the performance of obvious operations)" (p. 1). To face a problem means to realize a difference between the current state of affairs in the environment and another, in some way, specified state, between the actual state and the target state.

Problem solving, then, takes place when the reaction of the environment to possible human actions is not known from prior experience, but must be deduced by means of a mental representation of the causal or intentional structure and the functional state of the environment. These must be represented symbolically in a mental model as relationships among the elements of the model. The model can then serve to identify those human acts that will bring the causal elements into proper functional structure or initial conditions, or both. Intentional systems such as games and social systems will be discussed separately.

A major task in knowledge-based problem solving is to transfer those properties of the physical environment that are related to the perceived problem to a proper symbolic representation. The information observed in the environment is then perceived symbolically with reference to the conceptual framework available to the person. The environment is analyzed and broken down into elements at a level where the functional properties are known to the person and can be combined into a proper representation. How far this analysis need go depends on the person's familiarity with the situation.

In the present human-machine context, this analysis means an iterative consideration of the functional properties of the system at the various levels of representation in the abstraction hierarchy, between what is functionally needed and what is physically available. Elements in the representation at the various levels must be known from two points of view–from their functional value and the physically oriented implementation. Search for the proper model depends on identification top-down for elements "by means of which" or a bottom-up consideration of the "functional value" of the available objects and processes, to use Duncker's terms. Duncker (1945) describes problem solving as genealogic lines in a solution tree that very closely resembles iteration in the abstraction hierarchy used here, and discusses orderly top-down search as well as "suggestions from below," which are bottom-up considerations. He also realizes the importance of the guidance in the process that can be obtained from considering "avoidance of the evil" as additional demands upon the solution, a consideration that is very relevant for control of physical systems, in that reliability and safety requirements are very often stated as negative specifications. Alexander (1964) discusses in detail how an optimal design is best characterized by the consequences that are *not* acceptable.

In experimental problem solving, the actual state and the target state are often well defined or specified. In real-life problem solving, this is typically not the case. Very often, several goals must be considered and the priority depends upon the situation, and the actual state may be very obscure, e.g., in the case of failures in a technical system. These conditions will add to the difficulty of defining and establishing an adequate mental representation.

Formation of a proper representation of the causal structure of the environment depends on knowledge about the basic physical laws governing its behavior. This phase of problem solving is finished when a representation in a framework familiar to the person is obtained–which means a representation for which a set of rules for information processing is available. The representation then ceases to be an informative, symbolic framework and turns into a prescriptive system of signs that control the application of stereotyped process rules.

If the representation is then externalized in the form of a physical or graphic model, it is evident that the same kind of rule- and skill-based operation can be developed, as it is found for operation on the physical system itself. The efficiency of formal, mathematical models and technical graphs and diagrams, as, for example, control engineers' Bode plots and pole-zero graphs, depends on the existence of a large repertoire of stereotyped manipulation rules used for solutions and predictions–often to a degree where the fundamental understanding of the conceptual basis has decayed.

This in fact means that problem solving typically depends not only on knowledge-based transformations, but also on rule- and skill-based manipulations of a familiar model framework. The difference between these behavioral domains is mainly due to a difference in the basic control structure. At the rule-based level, the control is due to *rules* for actions upon the representation correlated to patterns in the *appearance* of the representation serving as signs. At the knowledge-based level, rules known to lead to a solution are not available; instead, control of operations depends on the symbolic interpretation of the representation and the *laws* representing the relationships among *symbolic elements*, such as the basic laws of nature. Then knowledge-based information processing is a goal- and concept-controlled search for proper rules for action, while rule-based processing will be the stereotyped use of such rules.

The conclusion of this discussion is that patterns in a model configuration, as well as perceptual patterns of the physical environment, can act as signs. This is most clearly seen if externalized representations of the mental model are actually available in the form of physical models, e.g., an abacus for calculation, or in the form of graphs or other symbolic representations on paper or on visual information displays, forming artificial objects for manipulation. For display formats designed for process control, this means that rule or skill-based control–"direct manipulation"–at a higher abstract level can be obtained if a symbolic display can be designed where there is a one-to-one mapping between the immediate appearance of the display and the properties of the process to be controlled.

In this way, the same conceptual model may act as a symbolic representation when considered in relation to the elements of the environment and the laws controlling their relationships, and as a system of prescriptive signs when considered in relation to the rules for model transformation and data processing. In complex situations, the correlation between cues from the representation and proper transformations may not be immediately obvious, but depends on a structured analysis of the cues–an analysis that may depend on logic or on higher-level stereotyped rules, like rules for systematic decision table entry (e.g., as found in naturalist guides based on Linné's taxonomy).

The distinction between signs and symbols in representations in the sense discussed above is equivalent to the distinction in semiotics and information

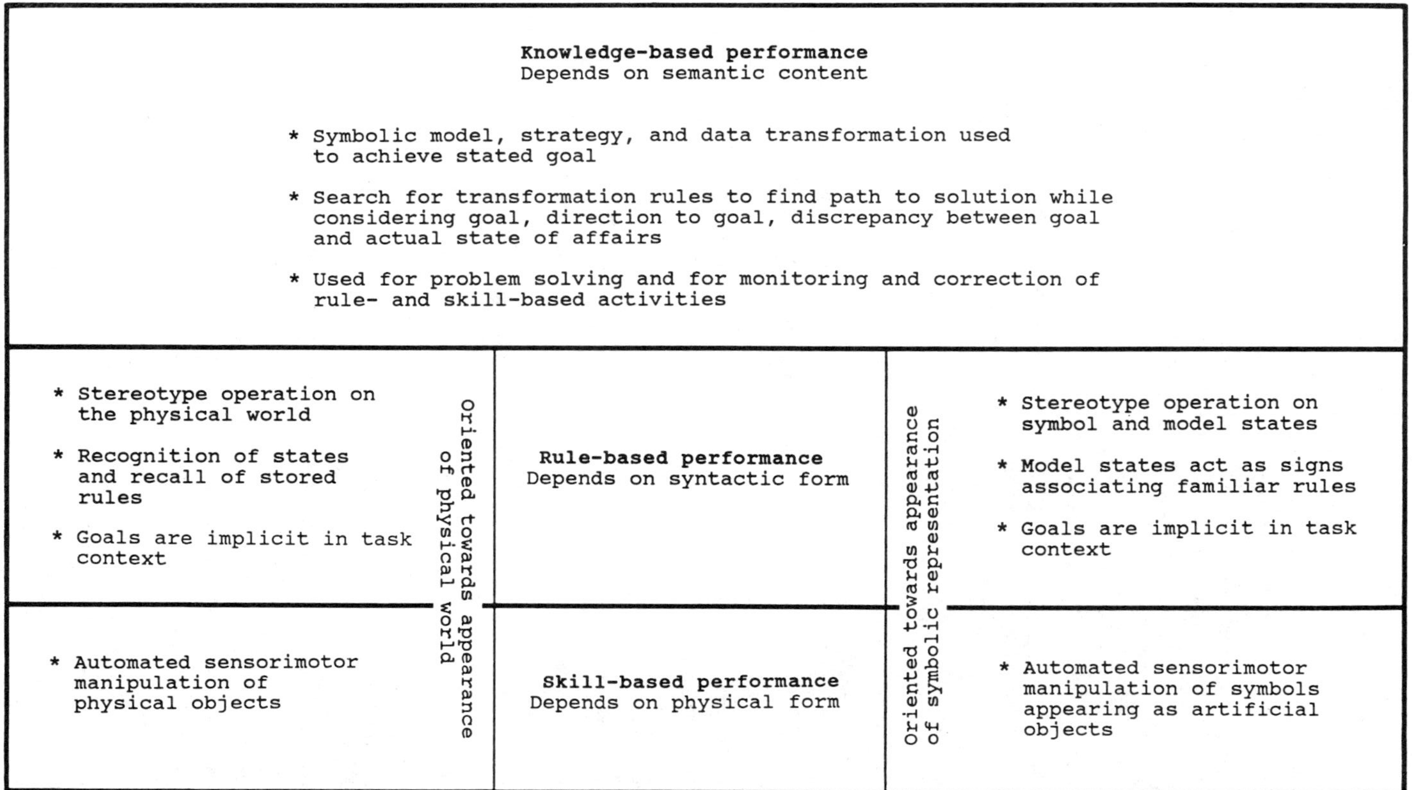

Figure 10.7. Skills, rules, and knowledge in a manual task and in problem solving.

science between prescriptive and informative texts (Morris, 1971; Eco, 1979). The important point in the present context is, however, related to the fact that the same text–or model–will be considered as prescriptive or informative by the same person, depending upon the situation. An important question is therefore how the cognitive level needed is activated and what information in fact serves this activation. This discussion closely relates to the philosophical discussion of the nature of mathematical propositions and their relation to the physical world (Gasking, 1953), i.e., whether such propositions, are for instance, well-founded empirical generalizations or they express rules for symbol manipulations. In the present context, the conclusion appears to be that their role entirely depends upon the situation of the user, his intentions, and background.

The role of a functional representation as prescriptive signs has been studied from a semiotic point of view by Cuny and Boy (1981). They analyzed the role of electrical circuit diagrams in terms of signs controlling activities during design, installation, and repair of electrical power supply systems in private houses, and investigated how different appearance of the same functional diagram was effective for support of the different activities.

In order to emphasize the role of rules and skill for symbol "manipulation" in problem solving, the three behavioral levels for operation upon a physical system, illustrated in Figure 9.1, can be extended as shown in Figure 10.7. When symbolic, graphic information displays are used for problem solving, the distinction between the two sides of the figure will disappear, and the high-capacity sensorimotor functions will be available for problem solving.

The Form of Mental Models

In the previous discussion of system representations in the abstraction hierarchy, only the *content* of the mental model, i.e., the system properties to be represented, has been mentioned. An important question is the *form* of the mental model, its structure, the elements, and relationships. We are here not so much interested here in the true description of some persons' mental models in some specific situation, but rather in a useful description of possible mental models that can be effective for persons in various tasks, and therefore useful for system design and evaluation. The form of mental representation must be inferred from the way humans express their knowledge about the functional properties of a particular system. People express such knowledge when they explain the functioning of a mechanism to other people during teaching and discussion, when they use functional reasoning during knowledge-based planning of a task, etc. The expressions are then available as verbal protocols or sketches and drawings.

A review of a couple of examples illustrates that at least two different forms of mental models must be considered. One is a mental model used in commonsense reasoning in terms of interacting objects; another is a more formal model in terms of variables and relations.

Figure 10.8. Typically, a mental model of a physical environment is a causal model structured in terms of objects with familiar functional properties. The objects interact in events, i.e., by state changes that propagate through the system. The purpose of or reason for a collection of objects can be so obvious that a model of physical form can turn into an animated behavioral model. The observer adds the reason and functional properties implied. [Reproduced with permission from Storm P. Museum, Copenhagen.]

The model behind commonsense reasoning is illustrated when the function of the system in Figure 10.8 is analyzed in order to explain it for another person. The structure of the system is represented as a spatial layout of the various components–in the form of the drawing itself or an internal cognitive map that may manifest itself by imagery. In the present case, the content of the model is the properties at the level of physical function, and the elements of the model will be physical objects, which are identified and separated from the general background; when the overall purpose of the arrangement has been realized, i.e., the objects are identified top-down in the abstraction hierarchy, even the explanation is performed only at the one level of physical function. In the mental model, each of the familiar objects has attributes representing their functional value, i.e., what they can be used for, and their functional properties, i.e., what they can do and what can be done to them by other objects. The function of the "system" can then be predicted or explained in the framework by means of if-then arguments resulting in a chain of events, i.e., changes of states

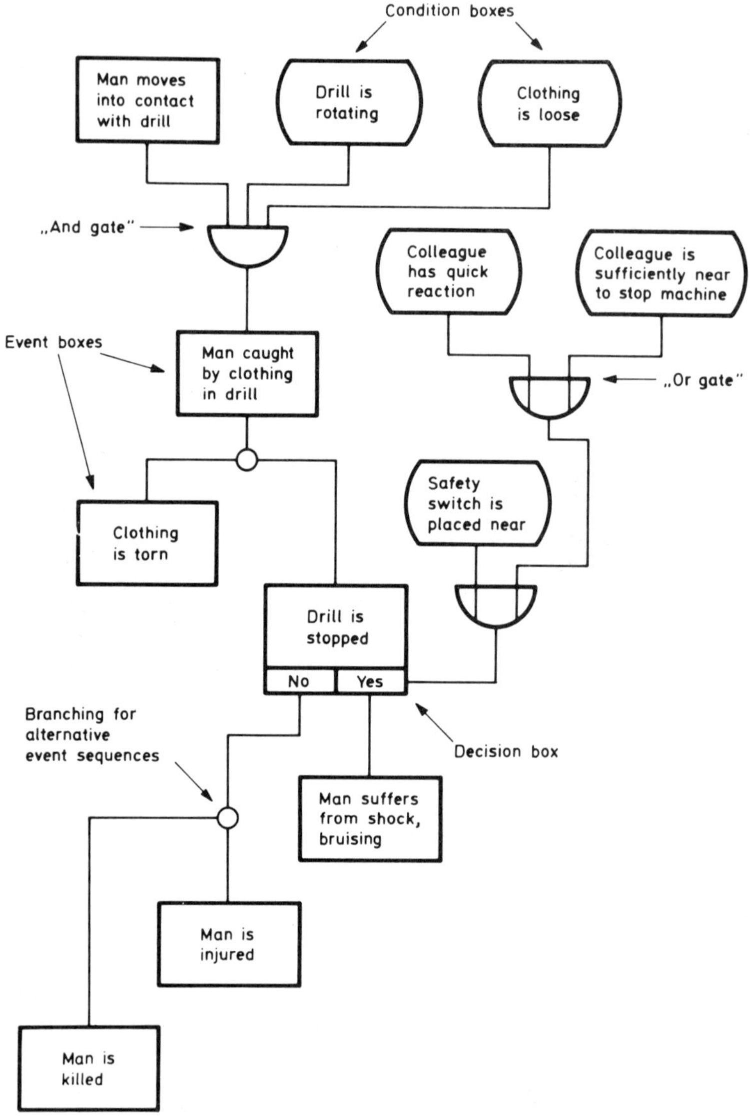

Figure 10.9. A cause-and-effect diagram represents the interrelation of potential events in a system. This is a stored set of if-then arguments derived from a functional model in terms of objects, properties and states, events. This is advantageous because of the close relation to physical events (faults).

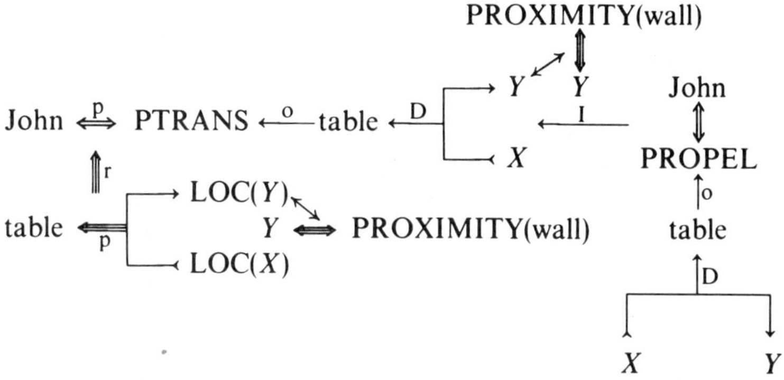

Figure 10.10. Semantic net representation of elements of natural language discourse–objects, events, states. "John pushed the table to the wall." [Reproduced from Shank (1975) with permission from Elsevier Science Publishing Co., Inc.]

by which objects interact. Such causal reasoning based on objects, states, and events is also used professionally for analysis of the propagation of events following a fault or error in a technical system (see Figure 10.9). The reason for this is that there is normally a close relationship between a fault or change in the real world and the corresponding change to update a model based on objects and states, whereas the changes necessary to update a formal deterministic model are more indirect and complex. Deterministic models appear to be well suited to describe the normal or intended, continuous function of a system, whereas the propagation of changes is more readily handled by causal models together with chains of if-then arguments.

The form of this mental model is close to the semantic nets of Schank (1975) and Rieger (1976) (Figure 10.10). Johnson-Laird (1980) also discusses internal representations in similar terms of a network of relations among objects. It should, however, be emphasized that this mental model only represents the structure that is used for conscious control of the higher-level strategy in reasoning. The total information process very much depends upon subconscious processes related to the internal dynamic world model that supplies the total context and thereby focuses the attention upon the appropriate phenomena and defines objects for consideration without conscious analysis. Without the context of the world model, the conscious verbal explanation loses precision unless extreme effort is spent on defining object properties in the general knowledge base and rules for selecting the relevant in particular situations. The present problem in AI appears to be related to the attempts to model human thinking by only representing the high-level verbal control and the lack of consideration of simultaneous use of several levels of abstraction.

Reference is frequently made to the ambiguity in analysis of isolated statements from commonsense reasoning. It is discussed in some detail by

$$m \frac{d^2}{dt^2}[s(t) + L \sin \phi(t)] = H(t),$$

$$m \frac{d^2}{dt^2}[L \cos \phi(t)] = V(t) - mg,$$

$$J \frac{d^2\phi(t)}{dt^2} = LV(t) \sin \phi(t) - LH(t) \cos \phi(t),$$

$$M \frac{d^2s(t)}{dt^2} = \mu(t) - H(t) - F \frac{ds(t)}{dt}$$

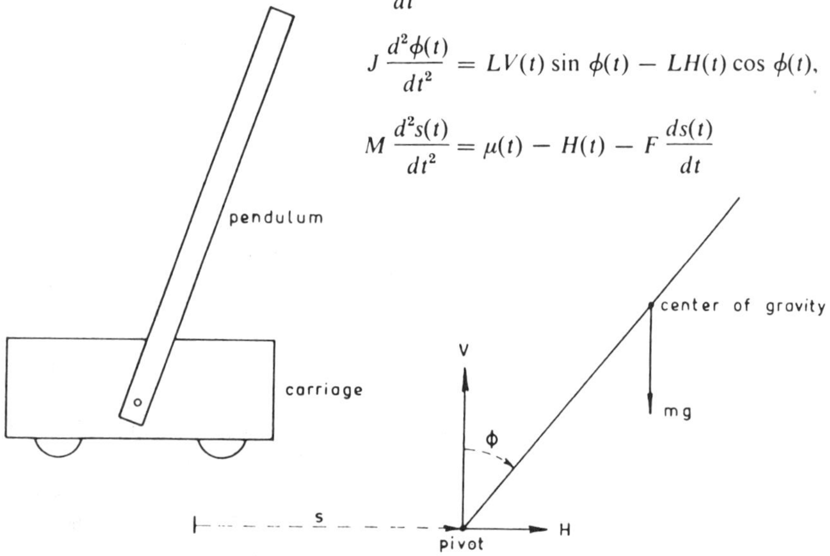

Figure 10.11. The scientific, deterministic representation of the physical environment is based on a network of relations among quantitative variables. This representation is complementary to the commonsense, causal model structured in objects. The example shows the representations of an inverted pendulum system.

Russell (1913), who distinguishes between commonsense, *causal* and formal, *deterministic* reasoning.

Deterministic reasoning is formal, scientific reasoning based on a mental model in terms of *formal relations* among well-defined quantitative representations of *physical variables*. In terms of the abstraction hierarchy, such models are typical for the level of generalized function. At the level of physical function, the models are related to sets of physical objects that are frequently met and therefore familiar. When physical, measurable variables are considered, it appears that certain sets of variables and typical relationships frequently appear and that formal representations of such generalized functions can be established independent of the particular physical configuration in the form of formal, mathematical relations between quantitative variables or derived representations such as graphs and tables. In the way that models based on recurrent familiar objects are well suited for commonsense reasoning, models based on recurrent relationships among physical variables will be well suited for formal, mathematical operations, because the formulas and rules for transformations are applicable for a large variety of physical contexts. A good example is a model of dynamic relationships in terms of differential equations based on Newton's laws (see Figure 10.11). Russell calls such models deterministic representations because the direction of causality is not represented in the set of equations. Any variables can be considered dependent variables and determined by calculation, given knowledge of the magnitudes of the remaining variables in the proper set of equations.

The two types of descriptions are in many respects complementary. In causal models, objects interact by events; the system is a set of objects related by a net of potential interactions in which changes or events propagate. Several quantitative variables are typically necessary to replace a description in terms of a state of a component or a mutual event between two components. The formal deterministic model is a network of relations among variables. In this model, the variables have replaced the physical objects as elements of the model. Physical objects are dissolved into a set of relations. To reflect a physical change of a component (owing to a fault or physical damage) in a formal deterministic model, a complicated updating of a set of relations is necessary. Updating of the qualitative, causal model to reflect a fault is much simpler in that the physical component is retained as an element in the model.

In the examples illustrated by Figures 10.8 and 10.11, the models also differ in the levels of abstraction. However, causal and deterministic models can be found at any levels of abstraction. A model of the dynamic properties of a specific mechanical arrangement in terms of relations among variables at the level of physical function can be derived from Newton's laws and other general laws by inserting the parameters and initial conditions in the proper set of equations. Or the model can be identified by trial and error in a search to match model predictions to recorded observations, e.g., a search that can be guided by aesthetic or metaphysical principles as was the case in the Ptolemaic epicycle models of planet motion. At the level of physical function, deterministic models are frequently represented in graphic form, i.e., relationships among variables are represented by graphs that can be derived from mathematical models or measured directly on a physical system.

On the other hand, commonsense causal reasoning is frequently used on higher levels of abstraction. A good example is a model of the generalized function of a feedback loop. An extensive mathematical modeling with well-developed procedures for calculation and transformations has been developed within the framework of deterministic models. However, for more informal and cursory reasoning, e.g., during systems design, an engineering practice for causal reasoning has developed. The "objects" of such reasoning are not physical objects, but abstract model concepts, such as loops, loop-gain, bandwidth, stability, overshoot, oscillation, etc. Causal reasoning then serves to track the propagation of design changes through the model structure. Very sophisticated tools have been developed to support such reasoning, e.g., in terms of pole-zero configuration charts that serve to investigate the effect of manipulations of gain-bandwidth characteristics upon loop performance (see Figure 10.12). In this way,

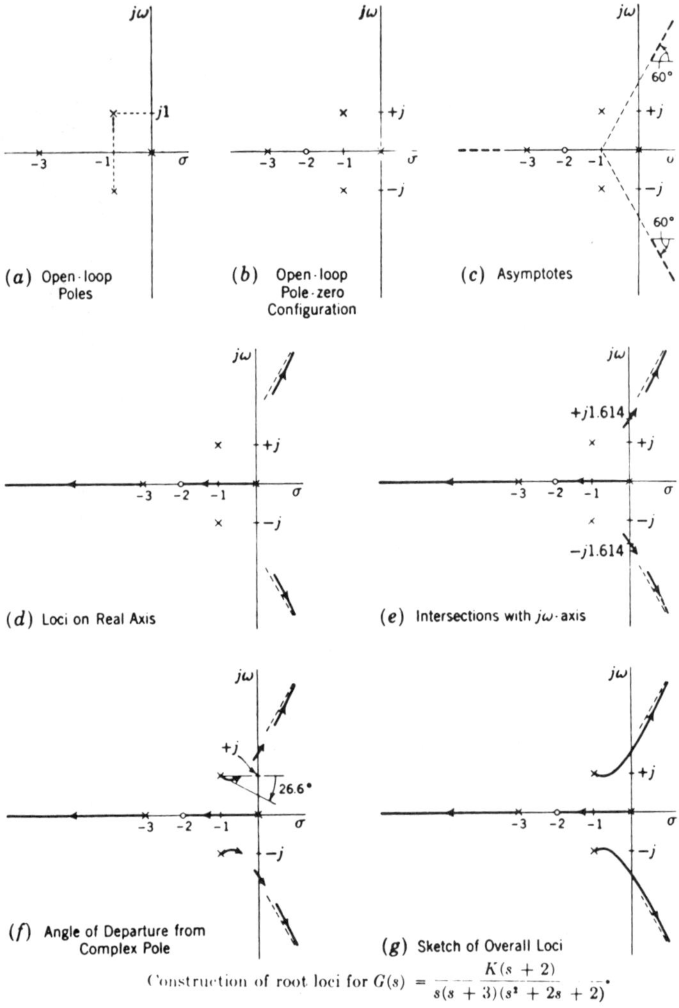

Figure 10.12. Rule-based symbol manipulation. Within control theory, the dynamic properties of a feed-back loop can be judged from the pole-zero-gain configuration and an extensive set of rules for manipulations of the configurations is developed for the control system design in this symbolic representation. The diagrams (a) to (f) show a set of rules for the graphic construction of the closed-loop root loci from the open-loop pole-zero configuration; (g) shows how the closed-loop poles and zeros of the transfer function of a feedback system move with increasing loop gain. [Reproduced from Truxal (1955) with permission from McGraw-Hill International Book Company.]

Figure 10.13. Elements of the mental model of a skilled operator as mentioned during start-up of a power plant boiler. The model is in terms of objects, states, and functional properties, here shown in relation to a mimic diagram of the underlying system. The statements are a literal translation of operators' short-hand terms. "Pump" is a label for coolant flow. "Cold slugs guzzle steam" is an objectizing of a general thermodynamic relation.

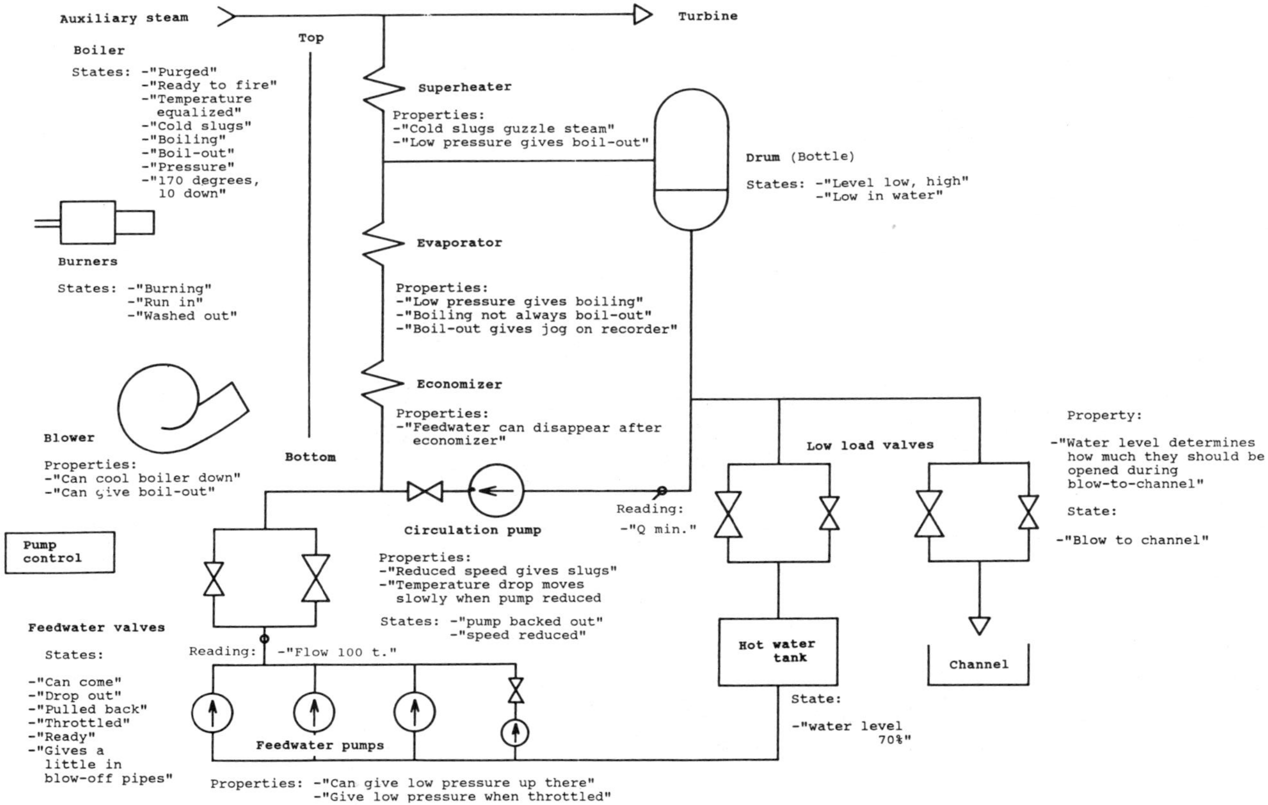
Auxiliary steam
Turbine
Top
Bottom
Boiler
States: -"Purged"
-"Ready to fire"
-"Temperature equalized"
-"Cold slugs"
-"Boiling"
-"Boil-out"
-"Pressure"
-"170 degrees, 10 down"
Burners
States: -"Burning"
-"Run in"
-"Washed out"
Blower
Properties:
-"Can cool boiler down"
-"Can give boil-out"
Pump control
Superheater
Properties:
-"Cold slugs guzzle steam"
-"Low pressure gives boil-out"
Drum (Bottle)
States: -"Level low, high"
-"Low in water"
Evaporator
Properties:
-"Low pressure gives boiling"
-"Boiling not always boil-out"
-"Boil-out gives jog on recorder"
Economizer
Properties:
-"Feedwater can disappear after economizer"
Circulation pump
Properties:
-"Reduced speed gives slugs"
-"Temperature drop moves slowly when pump reduced
States: -"pump backed out"
-"speed reduced"
Reading:
-"Q min."
Low load valves
Property:
-"Water level determines how much they should be opened during blow-to-channel"
State:
-"Blow to channel"
Channel
Hot water tank
State:
-"water level 70%"
Feedwater valves
States:
-"Can come"
-"Drop out"
-"Pulled back"
-"Throttled"
-"Ready"
-"Gives a little in blow-off pipes"
Reading: -"Flow 100 t."
Feedwater pumps
Properties: -"Can give low pressure up there"
-"Give low pressure when throttled"

Figure 10.13

the symbolic representations of any model may be reinterpreted as "artificial objects" and used in commonsense causal reasoning. Also in this way, causal commonsense reasoning is the forerunner of the formal, deterministic description in systems design, which often involves sequences of the arguments related to test of hypothetical changes in the design concept.

Because commonsense reasoning is well suited when elaborating on the propagation of changes through the physical environment, it will always be the basis of knowledge-based planning of the physical interaction with a system. Knowledge-based reasoning is most frequently called for when changes or abnormalities in the environment no longer allow behavior to be controlled by the lower rule- and skill-based levels. It is important to consider that this commonsense reasoning only serves to develop higher-level modifications of the lower-level behavior. The effects of causal arguments are determined by the context delivered by the subconscious dynamic world model, which will be updated during interaction and define the changes. Russell argues that causal reasoning is ambiguous and vague. This is, however, due to a consideration of the context-free, generalized verbal statements. Considered in the context, as higher-level control of behavior, verbal statements are very precise, which can be seen by the very specific effect of even rudimentary statements exchanged by people cooperating in system control (see Figure 10.13).

Chapter 11

Human Errors

Human errors and their prevention and prediction play an important role in human-machine system design. Furthermore, analysis of human errors is important for the attempt to validate models of human performance. Success in prediction and simulation of well-adapted human interaction is no proof of the validity of a model; only analysis of situations when adaptation breaks down will be informative, inasmuch as the aspects of human resources and their limitations must be reflected by a model. Models of well-adapted behavior largely reflect characteristics of the work environment.

Definition and Characteristics of Human Error

It is basically very difficult to give a satisfactory definition of human errors. Frequently they are identified after the fact: if a system performs less satisfactorily than it normally does–because of a human act or a disturbance that could have been counteracted by a reasonable human act–the cause will very likely be identified as a human error. When compared with technical components, human operators have some peculiar features that must be analyzed more closely to see whether the present general attitude toward faults and errors is reasonable and expedient. How are faults and errors defined? Faults and errors cannot be objectively defined by considering the performance of humans or equipment in isolation. They can only be defined with reference to human intentions or expectations; they depend upon somebody's judgment of the specific situation. Faults and errors are not only caused by changes in performance with respect to the normal or accepted performance, but also by changes of the criteria of judgments; i.e., changes in requirements to system performance, in safety requirements, or in legal conventions will be able to turn hitherto accepted performance into erroneous acts.

In the present human-machine context, we can define technical faults and human errors as causes of unfulfilled system purposes. If system performance is judged below the accepted, present standard, somebody will typically try to go back through the causal chain to find the causes. How far back to search is a rather open question; generally, the search will stop when one or more changes are found that are familiar and therefore acceptable as explanations, and to which something can be done for correction. In the case of a technical breakdown, a "component" failure is generally accepted as the cause at the component level where replacement is convenient. In some cases, however, component failure will not be found an acceptable cause: for example, if it occurs too frequently. In such cases, the search will often continue to find the "root cause" of the component's malfunction.

In summary, the characteristics of a fault are the following: it is the cause of deviation from a standard; it is found on the causal path backward from this effect; it is accepted as a familiar and therefore reasonable explanation; and a cure is known. In all these respects, the human operator is in an unlucky position. Because of human complexity, it is generally very difficult to "pass through" a person in causal explanations. In addition, it is generally accepted that "it is human to err," and finally, you can always ask people to "try harder." This means that allocation of causes to people or technical parts in the system is a purely pragmatic question regarding the stop rule applied for analysis after the fact. Ultimately, a thorough analysis will always end up with a human error, probably during design or manufacture, or with an act of God.

Errors as Unsuccessful Experiments in an Unkind Environment

A more fruitful point of view is to consider human errors as instances of human-machine or human-task mismatches. In case of systematic or frequent mismatches, the cause can then typically be considered a design error. Occasional mismatches are typically caused by variability on the part of the system or the person and can be considered system failures or human errors, respectively.

However, human variability is an important ingredient in adaptation and learning, and the ability to adapt to peculiarities in system performance and optimize interaction is the very reason for having people in a system. To optimize performance, to develop smooth and efficient skills, it is very important to have opportunities to "cut corners," to perform trial and error experiments, and in a way human errors can be considered as unsuccessful experiments with unacceptable consequences. Typically they are only classified as human errors because they are performed in an "unkind" work environment. An unkind work environment is then defined by the fact that it is not possible for a person to correct the effects of inappropriate variations in performance before they lead to unacceptable consequences. This is often the case because the person either cannot immediately observe the effects of "errors," or because the effects are irreversible. This is no new wisdom; as early as 1905, Ernst Mach said: "Knowledge and error flow from the same mental sources, only success can tell the one from the other" (Mach, 1905, p. 84). The interaction can be seen as a complex, multidimensional demand-resource matching process. To discuss the mismatches and evaluate means for improvement, it is more important to find the nature or dimensions of the mismatches than to identify their causes. In other words, it is necessary to find *what* went wrong rather than *why*.

Human Errors in Learning and Adaptation

The fact that human variability causes problems and, at the same time, is closely related to human learning and adaptation, deserves some closer consideration. Earlier, the change of the control of behavior during learning was mentioned; a key point was that during training, control structures at the lower cognitive levels are developed while, at the same time, higher-level structures can deteriorate.

The variation in human behavior when control moves downward during training and adaptation probably has important implications for human-task mismatches, which may ultimately be judged human error if not corrected in due time. In general, the only information available to the person to judge the proper limits of adaptation will be occasional mismatches of behavior and environment. In this way, conscious as well as subconscious experiments are part of the adaptation mechanisms at all levels of cognitive control.

The efficiency of human interaction with the environment at the skill-based level is due to a high degree of fine tuning of the sensorimotor schemata to the time-space features in the environment. Changes in the environment will often be met by an updating of the current schema by a subconscious reaction to cues or a consciously expressed intention ("Now look, be careful, the road is icy").

However, the updating of the current schema will frequently not take place until a mismatch has occurred, for instance, when walking onto more uneven ground, adaptation of the current motor schema to the actual features of the environment may first happen after the feet have detected the mismatch by stumbling. The point here is that adaptation and fine tuning of sensorimotor schemata basically depend upon mismatch occurrences for optimal adjustments. The proper limits for fine tuning can only be found if surpassed once in a while.

If the optimization criteria for manual skill development are the speed and smoothness of movements, optimization can only be constrained by the experience of the precision tolerance limits. This means that the shape of the distribution curve representing variability in time-space coordination is not a characteristic of the person's motor control, but reflects tolerance limits of the environment, and the "risk sensitivity" of the individual. This feature of human behavior has also been identified and discussed by researchers in traffic safety, which is related to a high-skill manual control task. It appears that beyond a certain limit, efforts to decrease accident frequency may influence the accident patterns, but not the general risk level (Taylor, 1981).

Also, the development of efficient rules-of-thumb and know-how at the rule-based level depends on a basic variability and experimentation to develop and adjust the proper rules, and to identify the information patterns that are suitable signs for controlling the rule application. The initial conditions for this adaptation by a novice are either knowledge-based rational planning or a set of simplified stereotyped procedures supplied by an instructor. In both cases, the process of adaptation will lead to experiments, some of which are bound to end up as human errors in unfriendly work environments.

The rational process of analysis, evaluation, and planning by an informed novice will not be maintained throughout a familiar work situation. The use of symbolic information for rational inference will gradually be replaced by use of convenient signs that are empirically correlated with the conditions necessary for the steps in a work procedure. This information may very well be informal information, such as relay clicks and mechanical noise. Such signs are, however, not reliable guides if the internal structure of the task environment changes, as it may in cases of component failure. In these situations, the convenient signs may lead the person into a trap in terms of acts based on wrong premises. Again, an occasional experience of unacceptable adaptation may serve basic control functions in the learning mechanism.

Formal work procedures will normally be based on signs and readings that are functionally defining the required initial states, and the planning of the steps and their mutual relationships in the work sequence will be made under consideration of likely variations in the work context. During adaptation, not only the formal sign will be replaced by more convenient, informal signs, but the sequence of work elements may, consciously or subconsciously, be rearranged to have a more natural and smooth sequence, judged from the immediate, normal experience with the task. This deviation from working "according to the rules" is the hallmark of experienced people, but is bound to give experiences that, depending on the consequences, give rise to human error and the related blame after the fact.

It should be considered here that the adaptation to informal signs and rules-of-thumb generally is not the result of conscious decisions, but is found as a result of the general variability of human behavior. Adaptation can be an evolutionary process, where effective variations survive and are integrated into behavior, whereas the unsuccesful are experienced as lapses and later avoided.

At the knowledge-based level where people are trying to cope with unfamiliar situations and therefore have to base behavior on functional analysis, evaluation, and planning, we will consider two major groups of human-task mismatches.

One group includes those cases in which people have proper intentions, but fail to implement them. In such cases, people may commit errors in reasoning because of, for instance, slips of memory, lack of knowledge, or to high workload–it may be difficult by unsupported, linear reasoning to deal with the complex causal net of the real world. It is not, however, possible to establish a complete set of preconditions for consideration in practical work situations and, logically, to make sure one's considerations are reliable. The only reliable test is to judge the response from the environment–and to correct oneself when unsuccessful, with the risk that one commits what later may be judged an error. Not even scientists are reliable–measurement of the atomic weights did only first converge on whole numbers when theoretical considerations asked for that and supplied the stop rule for the necessary efforts (Kuhn, 1962).

The other major category includes cases in which the humans' acts are in good correspondence with their intention, which, however, serves a subgoal not acceptable from an ultimate task or system performance point of view. An illustrating example may be the situation in which an operator in a disturbed process plant has several alternative hypotheses on the failed state that he has to test. The classic research on problem solving (Duncker, 1945) shows that it is a normal feature to develop several branches in a "solution tree" (see Figure 10.4) before one settles down for detailed consideration of one of the solutions. Theoretically, the test of hypotheses can be done conceptually, but faced with the system itself, testing by means of manipulations on the system will be a tempting solution. In case the hypothesis is incorrect, this act may add another disturbance to a system in an unknown state, and the result after the fact may be accusation of serious decision error.

The conclusion of this discussion is that "errors" are basically the effect of human variability in an unfriendly environment, and that this variability is an inherent element in human adaptation. In the following section, these aspects will be discussed in more detail in order to identify the most important classes of human error.

Error Recovery

If variation of human behavior is an important ingredient of development of smooth skills and professional know-how, and experiments on the environment are necessary for problem solving, definition of error should be related to a lack of recovery from unacceptable effects of exploratory behavior.

Error recovery depends on observability and reversibility of the emerging unacceptable effects. Reversibility largely depends on dynamics and linearity of system properties, whereas observability depends on the properties of the human-environment interface, which will be greatly influenced for many tasks by the use of advanced information technology.

Error observability depends on the perception of a mismatch between the expected and the actual system response to human actions. At the level of skilled behavior, the patterns of behavior are continuously adapted to the changes in the environment to absorb variations in coordination. Only when variations exceed the limits of adaptability in the current regime and cues call for modification, is it usually referred to as an error (for instance, stumbling). Error corrections then depend on the availability of alternative control patterns, and on the activation of these patterns by proper cues before control is irreversibly lost. This means that error recovery is tightly related to specific dynamic properies of the actual interface configuration.

In being related to the rule-based coordination of a sequence of skilled routines, error recovery may be influenced by various features. The information needed for control of actions and for observation of errors may be related to different time spans and to different levels of abstraction. The information used en route to control activity in pursuit of an intention or goal may be totally unrelated to the intention itself. In a habitual sequence of skilled action complexes, the individual complexes are released by stereotypical cues. Judgment of system responses in terms of intended outcome may require simultaneous functional evaluation at the knowledge-based cognitive level. Whether the knowledge basis required for this is maintained also for the more frequent tasks and the information necessary is available very much depends upon details in the human-task interface, in particular in tasks such as process control. This makes objective error data collection very difficult, because the bias caused by error recovery features of the various data sources is difficult to determine.

Mechanisms Behind Human-Machine Mismatches

For human-machine system design it is necessary to characterize and classify human errors in terms of human-system mismatches and the underlying cognitive control. Paradoxically, two major classes to consider include cases in which human variability is unacceptably great, and cases when human variability is insufficient to cope with changes in system performance. To explain human-system mismatch, we must therefore look at the control of human behavior, to find mechanisms behind variability during normal, familiar situations and mechanisms that limit capability in unfamiliar situations when the system changes.

In discussing these mechanisms, we are considering those occasions when a human-system interaction is judged a mismatch that needs correction–either by the person himself or by somebody else. We are not considering the success of the correction, i.e., the ultimate effect of the mismatch. Mismatches are typically corrected immediately by the person, but the success of the correction very much depends upon qualities of the task and environment, such as observability and reversibility, and must be discussed separately from the mechanisms behind the initial mismatch that led to a corrective action or adaptive change in behavior. See Figure 11.1.

Human Variability During Familiar Tasks

We will first consider intrinsic human variability, which leads to mismatches during normal work situations; i.e., we will consider the effect of variability upon skill- and rule-based behavior.

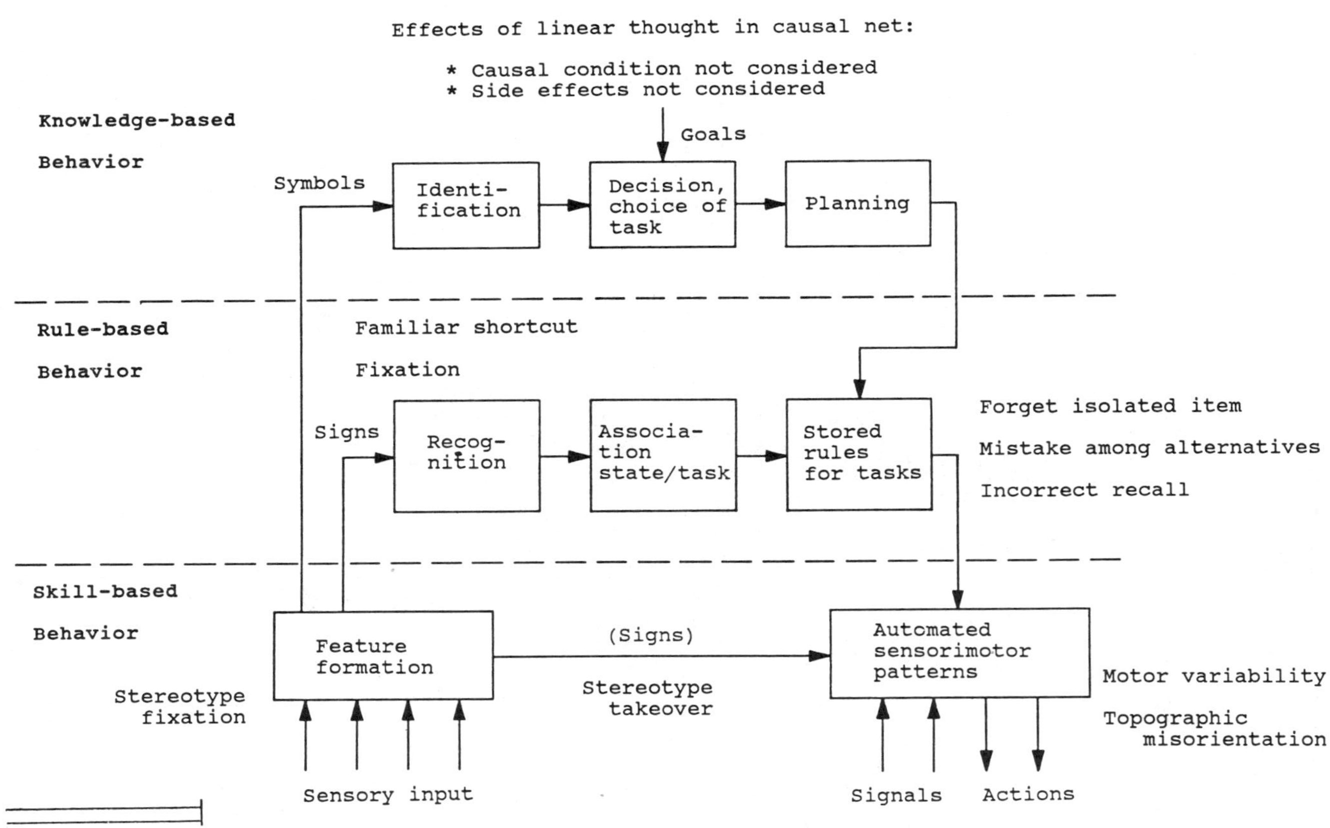

Figure 11.1. Typical human "error" mechanisms and their relations to control of behavior. [Adapted from Rasmussen (1980) with permission from John Wiley & Sons, Ltd.]

Motor variability. The time-space precision of skill-based sensorimotor control may not be adequate for the task at hand, leading to occasional mismatches. Examples are:

Inadequate precision leads to short circuit of terminals with screw driver.

Inadequate precision in replacement of relay cover leads to short circuit of relay terminals.

Varying use of force in manipulating a bank of valves occasionally leaves a valve leaking.

Topographic misorientation is another mechanism of mismatches during sensorimotor performance, occurring when the internal world model owing to some cause loses synchronism with the external world. For example: failure in one of several pump trains in the basement leads to the decision in the control room to switch off the "north train"; however, during passage downstairs, an operator loses orientation and switches off the southern train, even though he has the proper intention.

These are mechanisms within a single motor schema; other mechanisms are related to the fact that skilled operators have a large repertoire of schemata, and that a schema may involve a long sequence of acts. A single conscious statement of intention may activate a schema; whereafter, the attention may be directed toward planning of future activities or monitoring the past. The current, unmonitored schema will then be sensitive to interference leading to *stereotype takeover*. This means that another schema takes over the control, either because a part of the current action sequence is also part of another frequently used schema, or because of interfering intentions of the detached attention. Examples:

During normal operation of a process plant, the power supply to the instrumentation is disrupted. Investigation reveals that the manual main circuit breaker in the power supply is in the off position. The conclusion was that a roving operator, in checking cooling towers and pumps, had inadvertently switched from a routine check round to the Friday afternoon shutdown check round and turned off the supply. The routes of the two check rounds are the same, except that the operator is supposed to pass by the door of the generator room on the routine check, but to enter and turn off the supply on the shutdown checks. Something en route has obviously conditioned him for shutdown checks (sunshine and daydreams?). The operator was not aware of his action, but did not reject the explanation.

An experimental plant shuts down automatically during normal operation because of inadvertent manual operation of a cooling system shutoff valve. The valve control switch is placed behind the operating console, and so is the switch of a floodlight system, used for special operations monitored through closed-circuit television. The switches are neither similar nor closely positioned. The operator has to pass the valve switch on his way to the floodlight switch. In this case, the operator went behind the console to switch off the flood light, but operated the shutoff valves that caused plant shutdown through the interlock system.

During start-up of a process plant, the plant is automatically shut down during manual adjustment of a cooling system. During start-up, the operator monitored the temperature of the primary cooling system and controlled it by switching off and on the secondary cooling pumps to avoid water condensation in the primary system owing to the cold cooling water. On this occasion, he observed the temperature to pass the low limit, signaling a demand to switch off the secondary pumps, while he was talking to a cooperator about another matter over the phone. He then switched off the primary pumps and the plant immediately shut down automatically. He did not recognize the cause immediately, but had to diagnose the situation from the warning signals. The control keys for the two sets of pumps are positioned far apart on the console. However, a special routine exists, during which the operator switches the primary pumps on and off to allow an operator in the basement to adjust pump valves after pump overhaul while they communicate by phone. Is the event caused by schema interference resulting from the phone call?

Because the repertoire of automated sensorimotor schemata and their complexity increase with the skill that operators develop during their daily interaction with their system, the role of this kind of mismatch becomes more important with their experience, and can only be counteracted by making systems more tolerant of error.

At the rule-based level, human variability during performance of the normal, familiar tasks is most frequently found as an incorrect recall of rules and know-how. A characteristic category is *forgetting an isolated item*, i.e., what is not an integrated part of a larger memory structure. Typical is *omission of an isolated act* that is not a necessary part of the main task sequence.

The fact that the omitted steps are frequently unrelated to the verbal label of the task may be a condition directly contributing to their frequency. Analysis of industrial fires led Whorf (1956) to the conclusion that "the name of a situation affects behavior." Examples are abundant: "jumpers" not removed from terminals after repair; switches not turned back to "operation" after instrument calibration; bypass valves not reopened after pump repair, cables not reconnected after instrument repair, etc.

In an analysis of test and calibration reports from nuclear plants, it was found (Rasmussen, 1980a) that this category accounted for 50% of the analyzed reports. The high frequency can be due to high initial probability, but can also be due to the fact that the isolated acts are less likely to be observed and corrected immediately by the person.

A closely related mechanism is the incorrect recall of isolated items, such as quantitative figures, numbers, etc. Examples:

Incorrect recall of numbers of valves and switches.

Incorrect recall of figures, such as calibration references, set points, instrument readings.

Another frequent mechanism of variability during familiar tasks is the mistake among alternatives, which frequently appears as incorrect choice of one of a couple of possible alternatives to use, such as left-right, up-down, plus-minus, A-B, etc. Examples:

Using positive correction factors instead of negative; using increasing instead of decreasing signal in calibration.

Disconnect pump A instead of B.

These are all mismatch events caused by human variability in normal, familiar task situations. Other mechanisms lead to mismatch when humans fail to adequately adapt to variations and changes in the task environment.

Improper Human Adaptation to System Changes

The efficiency of human interaction with the environment at the skill-based level is due to a high degree of fine tuning of the sensorimotor schemata to the time-space features in the environment. Changes in the environment will often be met by an updating of the current schema by a subconscious reaction to cues or a consciously expressed intention ("Now look, be careful, the road is icy").

However, as mentioned above, the updating of the current schema will frequently not take place until a mismatch has occurred; for instance, when walking onto more uneven ground, adaptation of the current motor schema to the actual features of the environment may first happen after the feet have detected the mismatch by stumbling. The point here is that adaptation and fine tuning of sensorimotor schemata basically depend upon mismatch occurrences for optimal adjustments. The proper limits for fine tuning can be found only if surpassed once in a while. This means that mismatches cannot, and should not, be avoided, but a system must be tolerant and not respond irreversibly. This discussion relates to mismatches that are needed to control adaptation within the skill-based level. More serious mismatch categories are met when changes in the environment are not met by proper activation of higher-level control of behavior.

Two types of mismatch mechanisms are related to improper activation of rule-based control: stereotype fixation and stereotype takeover, similar to that discussed in the previous section.

Stereotype fixation represents the situation when a sensorimotor schema is activated in an improper context, and the person on afterthought very well knows what he should have done. He does not switch to proper rule-based control. Examples:

An operator presses air out of a plastic bag containing dust in order to seal it, although he knows it contains radioactive material. He gets contamination in his face.

During a cleanup operation in a radioactive area, a vacuum cleaner fails. A foreman opens it for a possible rapid repair, despite the fact that he knows it contains radioactive dust.

In both cases, normal everyday reactions are carried over to a special context. Also, this appears to be a reasonable and effective learning mechanism, in a reversible context. Here the invisible radioactivity makes the environment "unkind."

In other cases, people realize the need for the application of special procedures, but relapse to familiar routines; i.e., the stereotype takes over because of overlapping sequence elements. Examples:

You have noticed that the road is icy and decided to drive carefully, but when a dog enters the road you hit the brake.

An operator enters an emergency procedure and executes a sequence of actions correctly but then inappropriately stops a pump, an act that follows the sequence in another, more frequently used procedure, but here is wrong and risky.

An airplane is below acceptable altitude while approaching a runway. The pilot orders "full power" and the copilot responds correctly but also retracts the landing gear, resulting in a "wheels-up" landing. This act normally follows "full power" at low altitude during takeoff.

Subconscious control of sensorimotor sequences quite naturally has a high affinity to the very familiar routine sequences that are likely to "capture" the control (Norman, 1980). As we saw previously, this interference can happen between sets of familiar sequences, but is more probable in less frequent situations when conscious rule-based control is needed. This is particularly the case during situations in which the need for forward planning occupies the conscious attention as soon as the necessary rule has been rehearsed, i.e., before it has been executed.

Similar difficulties in proper adaptation to system changes by switching to knowledge-based behavior are caused by the reliance on signs during all familiar situations. The high efficiency of human interaction with objects and other persons of everyday life is due to a large repertoire of skilled subroutines and rules, and know-how for updating the routines and linking them together. The control is based on recognition of the state of affairs in the environment in terms of signs, which relates to the appropriate rules by convention or experience. Even in direct interaction with the physical environment will these signs be convenient correlates in the given context rather than defining attributes. This makes the interaction susceptible to mistakes if the environment changes in a way that does not affect the signs, but makes the related behavior inappropriate. This is basically the idea behind all kinds of hunting traps.

In the direct interaction with a physical world, identification of signs takes place by perceptual categorization, which can be based on complex patterns and therefore also be rather sensitive to changes. This is typically not the case for human interaction with complex industrial systems, where operators are controlling more or less invisible processes. They have to infer the state and select proper actions from a set of physical measurements that is seldom presented in a way that allows perceptual identification of the state; operators are supposed to apply conceptual categorization based on rational reasoning, i.e., to exhibit knowledge-based reasoning. For several reasons, this leads to difficulties for human operators to adapt appropriately to changes in the system, for instance, as caused by technical faults.

The use of a set of measured variables requires knowledge of the system in terms of engineers' conception as a network of quantitative relations among variables. Natural language reasoning, which is typically used for control of the systems, is, however, not based on nets of relations among variables, but upon linear sequences of events in a system of interacting components or functions. To circumvent the need for mental effort to derive states and events from the sets of variables and their relations, operators generally use indications that are typical for the normal events and states, including informal signals such as motor and relay noise, as convenient signs for familiar states in the system. This is a very effective and mentally economic strategy during normal and familiar periods, but leads the operator into traps in which changes in plant conditions are not adequately reflected in his set of signs. Such mental traps often contribute significantly to the operator's misidentification of unfamiliar, complex plant states. Therefore, to adapt performance to the requirements of a system in a unique and unfamiliar state, the operator must not only switch to knowledge-based reasoning based on a mental model of the internal, functional properties of the system, he must also replace his perception of the information as signs with an analytical interpretation as symbols. This appears to be very difficult, because the use of signs basically means that information from the system is not really observed, but is obtained by "asking questions" that are heavily biased by expectations based on a set of well-known situations.

The difficulty in shifting to higher-level analytical reasoning is further aggravated when inference must be based on a number of information sources that are *sequentially* attended. From analysis of verbal protocols recorded by skilled technicians in diagnostic tasks, we found several principles in operation that served to minimize memory workload. In reading a sequence of measurements, they did not try to remember the original observations; each reading was immediately judged according to their expectations and only the result of the judgment was later recalled. Furthermore, they followed a "way of least resistance" in that they made a decision about what to do next, as soon as a familiar approach seemed to be possible, without considering the possibility of alternative, more effective ways. Instead, there seemed to be a "point of no return," which had the effect that information observed after a decision would rarely lead to a reconsideration of the situation.

Taken together, these aspects force one to draw the conclusion that there is a considerable probability that highly skilled operators with a large repertoire of convenient signs and related know-how will not switch to analytical reasoning when required, if they find a familiar subset of data during their reading of instruments. They will instead run into a procedural trap and be caught by a *familiar association shortcut* very probably based on a subset of the available data. Examples from major industrial incidents are legion, but my favorite case story tells about Lalande, who failed to switch from his rule-based recording of star positions to an analytical interest for moving stars:

> The incident in question occurred in 1795, nine years after the discovery of the planet Uranus, and the principal figure involved was the great French astronomer Lalande. In that year Lalande failed to discover the planet Neptune, although the logic of events should have led him to it. Lalande was making a map of the heavens. Every night he would observe and record the stars in a small area, and on a following night would repeat the observations. Once, in a second mapping of a particular area, he found that the position of one star relative to others in that part of the map had shifted. Lalande was a good astronomer and knew that such a shift was unreasonable. He crossed out his first observation of the shifting point of light, put a question mark next to his second observation, and let the matter go. And so, not until half a century later did Neptune get added to the list of planets in the solar system. From the aberrant movement, Lalande might have made the inference not that an error had been made but that a new planet of the solar system was present. But he was reasonable. And it was more reasonable to infer that one had made an error in observation than that one had found a new planet. [From Bruner et al., 1956.]

The incident reveals that Lalande was not open-minded analyzing a physical system; he was absorbed in rule-based observation and data recording.

From a butadiene explosion in Texas City (Jarvis, 1971) the investigation considers:

> Loss of butadiene from the system through the leaking overhead line motor valve resulted in substantial changes in tray composition.... The loss of liquid in the base of the column uncovered the calandria tubes, allowing the tube wall temperature to approach the temperature of the heat supply. The increased vinylacetylene concentration and high tube wall temperature set the stage for the explosion which followed.... The make flow meter showed a continuous flow; however, the operator assumed that the meter was off calibration since the make motor valve was closed and the tracing of the chart was a straight line near the base of the chart. The column base level indicator showed a low level in the base of the column, but ample kettle vapor was being generated.

Given an unstable flow meter, only wisdom after the fact will make you consider a leak. This incident was, in fact, the basis of the example shown in Figure 9.3.

An example from the melt-down of fuel elements in a nuclear reactor shows the great affinity to familiar signs, even a prewarning has been received (Hanford incident, 1955):

> Certain tests required several hundred process coolant tubes to be blocked by neoprene disks. Seven disks were left in the system after the test, but were located by a test of the gauge system that monitors water presssure on each individual process tube. For some reason the gauge on one tube was overlooked, and it did not appear in a list of abnormal gauge readings prepared during the test. There was an additional opportunity to spot the blocked tube when a later test was performed on the system. This time the pressure for the tube definitely indicated a blocked tube. The shift supervisor failed, however, to recognize this indication of trouble. The gauge was adjusted at that time by an instrument mechanic to give a midscale reading which for that particular tube was false. This adjustment made it virtually certain that no flow condition would exist until serious damage resulted.

Once an operator has succeeded in shifting to analytical functional reasoning at the level of knowledge-based behavior, it is very difficult to characterize his mental data processing and the related mechanisms leading to mismatch.

At the skill- and rule-based levels, behavior is controlled by motor schemata and know-how rules, the goals are implicitly specified, and "error" mechanisms are described in terms related to established, "normal" action sequences in a rather behavioristic way. At the knowledge-based level, this is not possible. The sequence of arguments an operator will use during problem solving cannot be described in general terms; the goal to pursue must be explicitly considered, and the actual choice depends on very subjective and situation-dependent features. At the skill- and rule-based levels, it is known per definition that adaptation to changes is within the human capability in that we are considering familiar tasks. This is not the case when knowledge-based performance is required during complex disturbances. Therefore, different kinds of mismatch situations may occur:

Adaptation is outside the limits of capability, because of requirements for knowledge about system properties that is not available, or for data that are not presented; or because of excessive workload requirements.

Adaptation is possible, but unsuccessful because of inappropriate decisions, which result in acts upon the system, not in conformance with actual requirements. It must be noted here that these are only "errors" if they are not corrected timely; as discussed below, actions not conforming with system requirements can be an important element during problem solving.

In the present context, only the latter category is considered, and again different typical categories of mismatch situations can be found during any of the necessary phases of the decision making, such as identification of system state, evaluations and choice of ultimate goals, and planning of proper action sequence:

Human variability in a cognitive task, such as slips of memory, mistakes, interference from familiar lines of reasoning, etc. These mechanisms are similar to those discussed above. They are difficult to identify or to use in prediction, when the problem-solving process is as unconstrained as it is in a real-life task in a control room.

Errors caused by the difficulty of keeping track of sequential reasoning in a causal structure, which is in fact a complex network, unsuited for linear reasoning. The mental workload involved may lead to adoption of premature hypotheses from the influence of factors as the way of least resistance and the point of no return, leading to *lack of consideration of important conditions or unacceptable side effects* of the ultimate decision.

Actions not conforming with system requirements may not be related to the ultimate result of erroneous decision making, but a reasonable act to *test a hypothesis* or get information which, however, may bring the system into a more complex and less controllable state.

The need for human decision making during disturbed system conditions basically depends on functional redundancy in the purpose/function/equipment relationships of the system. There is a complex many-to-many mapping between the levels in this hierarchy, and during search for resources to resolve the various goals in a complex situation, operators may very likely be caught by familiar or proceduralized relationships serving goals that are not relevant to the present situation. Decision errors during complex disturbances are not stochastic events, but very often reasonable mistakes caused by interference in this mapping.

The conclusion is that in present-day control rooms based on individual presentation of the measured variables, the context in which operators make decisions at the knowledge-based level is far too unstructured to allow the development of a model of their problem-solving process, and hence to identify typical error modes, except in very general terms, such as "lack of consideration of latent conditions or side effects" (Rasmussen, 1980a).

As a basis for a useful model, the conceptual framework within which the operators have to make decision has to be modeled in a consistent way in terms of the purpose/function/equipment hierarchy of the system. As Simon notes (Simon, 1969): "the complexity of human behavior largely reflects the complexity of his environment...." Before his behavior can be modeled, a systematic description of his decision making context must be found in order to identify likely interferences. Second, realistic models will probably only be possible if his choice of goals and strategies is more constrained and controlled than is the case in present-day control rooms. This is possible if a computer-controlled presentation of a symbolic framework is developed, which may lead to skill- and rule-based problem solving in an externalized context.

Taxonomy of Human Errors

In order to characterize human errors as human-machine mismatch situations, the categories of psychological mechanisms discussed in the previous section must be supplemented by a description of the interaction with the environment, the machine. The psychological mechanism described is only one element in the causal sequence we meet during causal backtracking from the occurrence of unacceptable system performance.

The first element of the mismatch we meet during backtracking will be an unacceptable state in a physical component owing to a human act, i.e., features of the mismatch in terms of inappropriate task performance. This dimension could be called the *external mode of human malfunction*, in order to avoid the term "human error," which has a flavor of guilt. In this category, mismatch is expressed in terms of omission of acts in a procedural sequence, wrong timing, etc. In order to relate these elements to the psychological mechanisms, a cognitive task analysis should be applied to identify the *internal mode of malfunction*, i.e., which element of the internal cognitive control was affected, either by not being properly performed or by being improperly bypassed. In practice, this is possible in terms of the elements of the decision sequence of Figure 2.1, which in acceptable detail will take care of rule- and skill-based behavior during analysis of incident reports. The strategies of knowledge-based behavior can only be considered by very careful analysis in cooperation with the actual people (e.g., see Pew et al., 1981). This is the reason for the very rudimentary discussion of knowledge-based mechanisms in the previous section, which was based on analysis of routine event reports from nuclear power plants. In consequence, the elements in the category of internal malfunction directly reflect the steps in the "decision ladder." Frequently, the mismatch is not due to spontaneous, inherent human variability, but events in the environment, which act as precursors, can be identified. Such events can be causes of the human malfunction, and the causal backtracking can be continued upstream from the human. In the present context, only events recognizable at distinct locations in time are considered to be causes, such as, for instance, interference by communication from colleagues, telephone calls, events in the physical environment, and bursts of noise. More persistent conditions, such as bad ergonomic design, generally high noise levels, inadequate instruction, etc., are considered *performance-affecting factors*, which do not release malfunction but change their likelihood. These factors include the influence upon performance through the psychological and physiological levels below the level of information processing in the modeling hierarchy of Figure 8.1. Supplemented by a representation of the task content, these different features of a human-system mismatch add up to a *taxonomy* that is useful for collection of human error data and for analysis during system design (see Figure 11.2).

This taxonomy does not lead to a generic, hierarchical classification system, but to a multifacet system to characterize mismatch occurrences. This approach

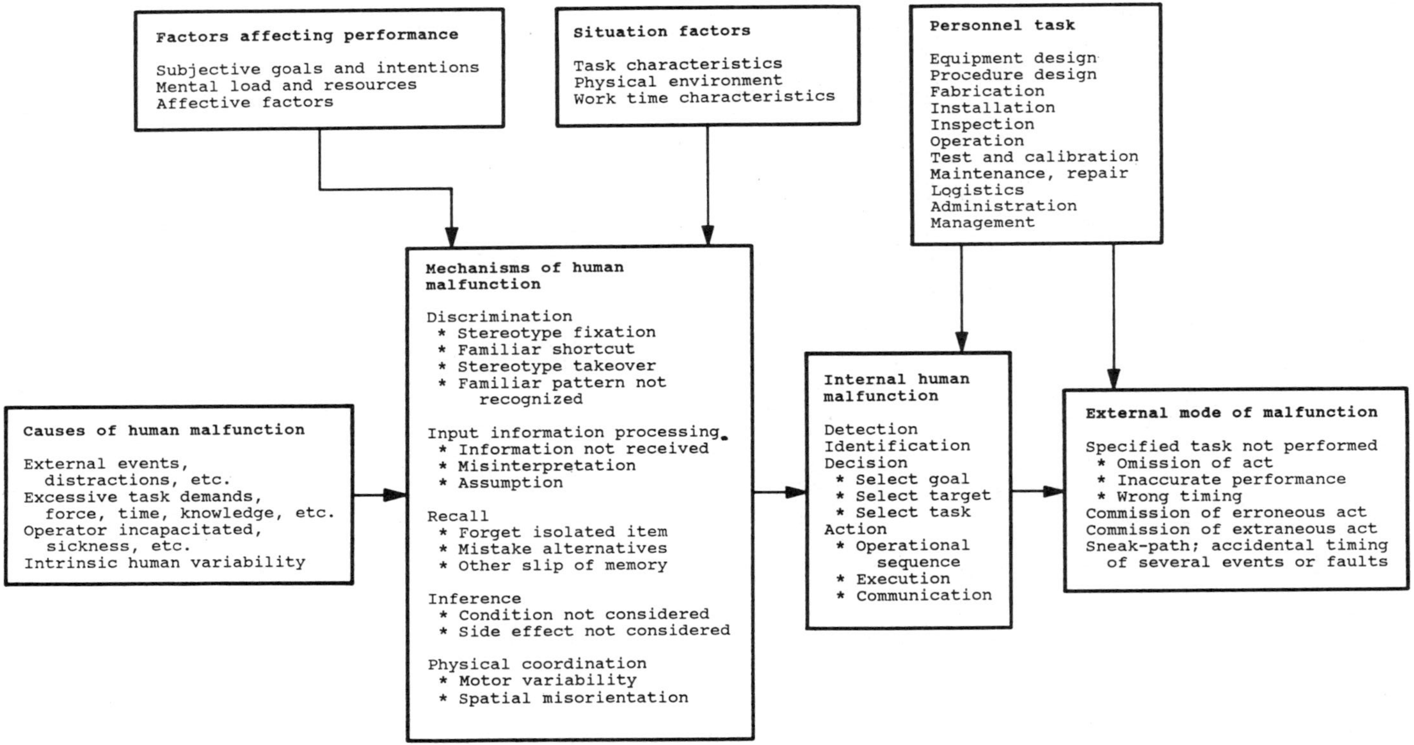

Figure 11.2. Multifaceted taxonomy for description and analysis of events involving human malfunction. [Reproduced from Rasmussen (1982) with permission from John Wiley & Sons, Ltd.]

has several advantages. First of all, a very high resolution in the description can be obtained with only a manageable few elements in each of the dimensions of the taxonomy. Furthermore, the internal structure of an event is preserved in the description, and can be regenerated in another context for analysis during design of new systems. For instance, very complex, but likely, scenarios for mismatch events can be identified and studied by postulating an internal mismatch mechanism, which is then folded into the different decision phases under the assumption of level of training of the person. The resulting relevant "internal malfunctions" are then correlated with each of the steps in an assumed work procedure by careful consideration of potential interference in the decision context, and the combined effects of possible inappropriate acts are assessed for different alternative system configurations. The probability of the significant courses of events could then be judged by identification of the likely causes and the quality of work situation in terms of performance-affecting factors. The advantage is that complex scenarios can be directly identified from the underlying psychological mechanisms. If the analysis is based only on the consideration of human errors in terms of their external manifestations as, for example, omissions, commissions, and inappropriate timing, the search for complex and risky scenarios will be hindered by a combinatorial explosion (Rasmussen, 1982a). The multifacet taxonomy of Figure 11.2, supplemented by categories to characterize the technical system and component faults, has been considered for an event reporting system for nuclear installations by an expert group under the OECD committee on the safety of nuclear installations (Rasmussen et al., 1981).

Human Information Processing and Stress

In the discussion of human error mechanisms, it has thus far been assumed that the normal cognitive processes are active, and that the general factors in the work situation affecting the performance would do this by changing the probability of error, rather than changing the organization of the cognitive performance drastically; i.e., only moderate levels of emotional stress are considered. In terms of the levels of modeling in Figure 8.2, this means that the influence from the emotional and affective psychological processes does not bring the cognitive performance outside the domain that can be described by the information-processing mechanisms identified during near-normal work conditions.

However, for analysis of human performance in supervisory control tasks, an important problem is to model how stress developing in situations with high workload, threat on the individual, or debilitating physical environments, e.g., noise, affects the human information processes and, in particular, the error characteristics. Unfortunately, the cognitive approach is not typical of stress research. The information available is typically from either behavioristic experiments relating stressors, e.g., noise level, to reaction times, or is very general statements as the uncalibrated inverted U-shape relation between stress and performance, or as statements describing "tunnel vision" in cognitive activities in levels of high stress. In addition, because of the lack of cognitive analysis, the fact is typically not considered that the cognitive task itself will normally change during situations of high stress due to the unfamiliarity of the situation.

Thus, two factors need to be analyzed. A theory relating the influence of stress and other affective factors upon the basic organization of the human information processes should be developed. This is, however, a problem of basic psychological research and will only be very briefly discussed here. Furthermore, the influence directly on the choice among the normally available information processing strategies should be described. We will consider the latter factor first, because it can be described within the present modeling framework.

In the human-machine context, periods inducing stress are generally periods of major disturbances of operation when high risk for losses is present together with time stress caused by system dynamics. Independent of emotional stress, the cognitive task will be changed in a way that leads to tunnel vision or narrowing of attention from at least two causes. First, the identification of an unusual state in the physical system depends on functional reasoning, i.e., diagnosis through hypotheses and test or topographical search strategies, which require much higher information capacities than is the case during familiar periods when identification is based on recognition of signs. This directly implies that less capacity is available for general monitoring and scanning to maintain a wide field of attention. Second, during normal periods a wide field of attention can be maintained because monitoring and observational scanning can be performed at a high functional level, in that variables related to the same functional unit will be highly correlated, and only representative indications for each unit need to be included in observation scanning. In case of disturbances, however, the correlation of variables from the disturbed function changes and, in order to diagnose, the entire set of variables from that function must be included in the observation. Consequently, the observer will focus on the disturbed function with a "cognitive tunnel effect" as the result (Moray, 1981).

Also, the disintegration of behavior during periods of high stress may be directly related to the structure of human information processing with its frequent shift in strategy. In general, an information-processing task may be based on various strategies with different resource/demand characteristics, and when difficulties are met during a task, a shift in strategy will probably be made in order to find a better resource/demand fit; see Chapter 7. This tendency may be particularly pronounced during high risk or time stress and therefore lead to a hectic looking for a better way of approaching the problem.

Another consideration is the influence upon performance directly from the affective arousal involved in stress. Unfortunately, no psychological model is available to relate affective stress to the parameters of the processing mechanisms. In his review, Rabbit (1979) states: "No current models in psychology specify how the functional mechanisms underlying human performance at any simple task may change their characteristics." However, the approach by Broadbent (1971) is an attempt in this direction. Broadbent's model of human information processing identifies three distinct mechanisms: perceptual encoding, translation processes, and response selection and execution. The assumption is then that any stressor, or general condition of the system, may potentially affect the operation of any stage but not others. From results obtained from laboratory experiments applying stressors such as noise, heat, alcohol, and loss of sleep, Broadbent supposes that all stressors affect perceptual selection and transformation, but that only stressors that lower arousal, e.g., lack of sleep, affect response selection. Other approaches to modeling the influence of stress upon elements in an information process model have been based on Sternberg's (1969) paradigm. So far, the results of laboratory experiments have been difficult to transfer to real-task performance; for a review, see Rabbit (1979).

What is really needed at present seems to be experiments in real tasks, for instance, in training simulators. A classic experiment is described by Bartlett (1943), who analyzes the influence of fatigue upon pilot performance. Bartlett argues that the classic experimental technique to analyze degradation of an elementary task from fatigue when repeated through time has no bearing on real-task performance, because real tasks are characterized by the demand for coordination of several elementary activities. This position is supported by the analysis of the degradation of skill from fatigue through time of flying. The result is that skilled subroutines do not deteriorate, but the higher-level coordination does. When fresh, the operator perceives instrument indications as a whole and relates control actions to overall performance of the aircraft. Fatigue leads to a dissociation of the "stimulus field" and of the actions into separate elements. An action was not performed to control the aircraft but to correct an instrument indication. Furthermore, subtasks of occasional nature, such as fuel replenishment, are omitted, and sensory information of secondary nature, such as bodily sensations that a fresh pilot integrates into the overall pattern and uses skillfully, becomes disintegrated and distracting. It appears that analysis of the influence of stress upon higher-level information processing functions is highly desirable.

Also, Mandler (1975, 1979) relates stress to conscious, cognitive processes and to interruption of activity. He applies a cognitive definition of stress: "External stressors are effective to the extent that they are perceived as dangerous or threatening, that is to the extent they are cognitively interpreted." According to this view, the most important effect of stressors on thought processes is that they interfere with consciousness and thus with the smooth operation of cognitive processes. Mandler views the human organism as preprogrammed to represent certain events consciously. Among these are intense stimuli and—important for his definition of stress—internal physiologic events and autonomic neural activity. Whenever such events call for conscious attention, the limited capacity will be drained, and other cognitive functions will deteriorate. His basic premise is then that autonomic activity results whenever some organized action or thought process is interrupted. This may occur because of some blocking by external events, or because some internal thought process prevents the completion of another process.

An interruption can occur in perceptual, cognitive, or behavioral domains, and in all cases the consequence will be autonomic activity. The consequence of the definition of stress as an emergency-signaling interruption and competition for capacity of cognitive mechanisms will be that it can have the effect of increasing attention to crucial events in the environment and more efficient attack on central problems. This is not only mentioned by Mandler (1979) but also by Baddeley (1972), quoted by Mandler: "A dangerous situation will tend to increase the level of arousal which in turn will focus the subject's attention more narrowly on those aspects of the situation he considers most important. If the task he is performing is regarded by him as most important, then performance will tend to improve. On the other hand, if it is regarded as peripheral to some other activity, such as avoiding danger, then danger will deteriorate." The position taken by Mandler is very much in line with the mechanism discussed in the beginning of this section.

The emphasis of the above discussion of models of stress has been to indicate that unusual situations influence the control of attention and the choice of strategy even though the phenomenon called stress is not present, and to cite references to approaches to models of stress that are compatible with information-processing models of human behavior considered for systems design. For more general treatment, readers are referred to sources from social and psychological research (for instance, see Janis and Mann, 1977). This literature is important to consider for analysis of human errors, when situations with high degrees of stress may bring operators and system users outside the domain of behavior for which the models considered in this text are applicable.

Chapter 12

A Catalogue of Models

The skill-, rule-, and knowledge-based framework is not a predictive model of human behavior, but rather a taxonomy that can be used to characterize various categories of behavior and to distinguish among them. Because the control mechanisms of the different categories, as well as the interpretation of information from the environment, are basically different, it is likely that different types of models are suitable for quantitative prediction in the various domains. Analytical models can be very important in the design of human-machine systems for optimizing task performance, in particular, for rare event scenarios that are to be considered for design in centralized, high-risk installations and for which system performance cannot be verified experimentally. A problem one has to face in such situations is that modeling of the interaction of categories of behavior is very important, and this requires compatibility among the separate kinds of models. It is not intended here to go into detail on the various models, but instead to review their main characteristics and to interrelate them and discuss the boundaries of their usefulness in the light of the skill-rule-knowledge framework.

The natural starting point of this discussion is a review of the approaches toward the modeling of the subconscious, skilled interaction with the physical environment. Behavior in this domain has some characteristics that determine the nature of useful models. Behavior is not controlled by a set of process rules, but by a dynamic, internal model of the environment that controls the sensorimotor interaction by means of signal processing in the space-time domain. Because skilled performance only includes behavior of well-adapted persons, models of performance in this domain will generally be models of features of the environment as viewed through human selectivity and limitations. If models should also include the quantitative precision necessary in control of manual acts, sets of mathematical time functions seem to be the only reasonable choice.

Models in this category have been discussed in detail by Sheridan and Ferrell (1974). Two aspects of skilled behavior are to be considered, and have typically been considered separately, in modeling. One is the control of human attention allocation; the other aspect is the control of the manual interaction.

Attention Allocation

Humans do not constantly scan the environment and extract meaningful features from the available flux of information. Acting in a familiar environment, people sample the environment, controlled by their expectations as to where an update of their internal model is needed. This means that this model specifies when an update is needed, and where to look. Different approaches to model this function have been tested experimentally.

One family of models is based on *queueing theory*. The system considered in queueing theory is a person serving a number of tasks. The tasks cannot be attended simultaneously, but have to be considered on a time-sharing basis according to a service strategy depending on the nature of the tasks. Many task demands, such as instrument reading during a monitoring task, arrive randomly. Typically, queueing theory considers demands with Poisson or exponential distributions. Queueing models of attention allocation postulate that humans optimize their performance according to a service strategy considering the arrival sequence and task priority. Queueing theoretic models have been used by Carbonell (1966) and Carbonell et al. (1968) for a study of instrument scanning behavior of aircraft pilots in order to predict the fraction of time devoted to each instrument. Also, Senders and Posner (1976) have developed a queueing model for monitoring tasks. Rouse (1977) has employed queueing theory to model pilot decision making in a multitask flight management task. Queueing models basically represent the time distribution and priority characteristics of the task environment and can therefore be useful for analysis of the workload posed in terms of time and scanning requirements in a monitoring task.

Another approach in the frequency domain is based on Nyquist's information *sampling theorem*, which states that the information from a source having spectral components with an upper limit frequency of w Hertz can be completely represented by an observer who samples $2w$ times per second. The sampling model has been tested by Senders (1964) in experiments where the subjects' task was to respond to a number of instruments fed by random signals of different bandwidth. Also, data from pilots in real flight tasks seem to match the model (Senders, 1966).

Detection

The ability of humans to detect changes in the behavior of a system that they are monitoring is important for systems control, and different approaches to predictive, quantitative models have been taken.

A classic approach is based on the *signal-detection theory* that was developed from studies of problems in statistical analysis of detection of radar signals in noise (Wald, 1947). A review of later developments can be found in Sheridan and Ferrell (1974). An extensive number of experiments have been made with subjects detecting visual or auditory signals in a background of noise. Given probability distributions of noise and signal plus noise, signal-detection theory can lead to prediction of the probability of misses and false alarms in a detection task. The theory has, however, had more application in experimental psychophysics than in systems design.

Detection of changes in system behavior from signals, rather than appearance of signals in noise, has been modeled by *estimation theory*. This is a control theoretic approach based on parameter estimation in a state-space representation. Models based on estimation theory represent human detectors as ideal observers monitoring dynamic processes. Gai and Curry (1976a) have proposed such a model, which assumes that the human observer can be represented by an optimal Kalman filter that serves to eliminate perceptual noise and to predict the state of the system monitored. Detection is, then, based on the Kalman filter residuals, i.e., the difference between the observers' expectations regarding the states and the actually observed states. In this type of detection model, only visual input information is considered, i.e., the system is an open loop. It has been suggested, partly based on Young's (1969) experiments, that feedback from the hand movement of a closed-loop human controller will improve failure detection (Curry and Ephrath, 1976). Subsequent experiments tend to indicate that the influence on detection of closed-loop control depends on circumstances. A series of experiments made by Wickens and Kessel (1981) confirm the hypothesis of improved closed-loop detection, whereas Ephrath and Young (1981) report ambiguous results possibly depending on the level of workload in the task. Another explanation may be that perception and motor schema generation have different characteristics and limit properties in their functions of dynamic world modeling. Probably, comparative experiments with detection in open-loop monitoring and closed-loop control could be used to separate aspects of these functions.

The attention allocation and detection functions basically depend on signal processing. The information considered is the quantitative state of signal patterns related to spatially distributed sources with reference to their stochastic nature. For these functions, the functional meaning with respect to selection of proper action is not considered. This interpretation is represented separately in judgment- or decision-making models.

Manual Control Models

Models of human performance in closed-loop control tasks have been very important for representation of the properties and limitations of humans in

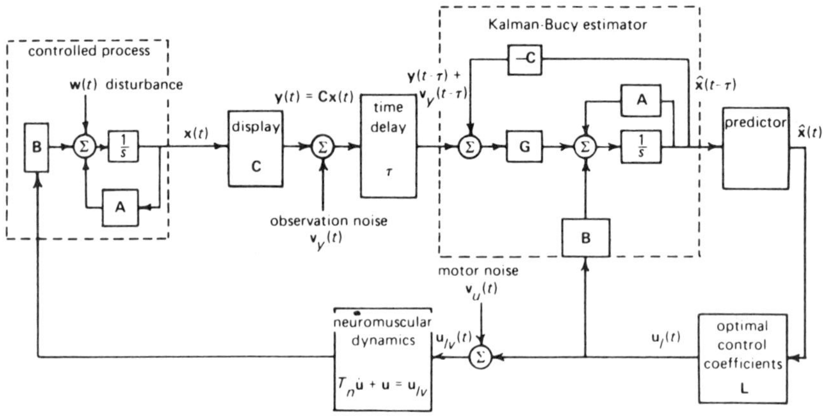

Figure 12.1. A typical optimal control model of a human vehicle controller. [From Kleinman et al. (1971) with permission from IEEE.]

vehicle control, in particular in aviation, and a separate school of modeling based on control theoretic tools has developed.

A review of models of human sensorimotor performance has been given by Pew (1974). Such models have been developed, in particular, from laboratory tracking tasks, and will typically give information on signal-to-noise ratio, maximum bandwidth when tracking unpredictable wave forms, prediction capabilities in sine-wave tracking, etc. The role of predictive feed-forward control for an industrial control task has been demonstrated experimentally by Crossman and Cooke (1962). For design and evaluation of the control system for aircraft, for example, models based on continuous linear differential equations are important because they are suitable for computer simulation of the total system performance. A review of such models can be found in Sheridan and Ferrell (1974).

The most recent development of such models has led to the optimal control models, which are based on the observation of Leonard (1960) and Roig (1962) that the mean-square error from human tracking data approximated the mean-square error of various optimal controllers. Optimal control models have in particular been developed by Baron and Kleinman (1969) and used for describing pilot performance (Kleinman et al., 1971). This model is based on the assumption that a well-trained, well-motivated human operator will act in a near-optimal way subject to certain internal constraints that limit his behavior. The internal dynamic world model necessary to account for human anticipation is represented by a Kalman-Bucy optimal filter. The model also includes observation noise and time delays depending on the instrument scanning strategy. The criterion used is typically minimum square deviations from desired output as well as squared control effort according to a chosen trade-off ratio. See Figure 12.1.

The model has been developed beyond the simple man-in-a-servoloop case, in that higher-level sequential functions for parameter and criterion control have been added in order to include multivariable control, monitoring, and decision making. This effort has proved successful for flight control and landing approach planning (Muralidharan and Baron, 1980; Baron et al., 1981). Efforts have also been made to extend this model to process control, and recently (Baron et al., 1982), a simulation model based on control theory has been proposed for simulation of the dynamic performance of a nuclear power plant including the operating staff. One important aspect of this approach is that human behavior at all three levels (skill-, rule-, and knowledge-based) as well as their interactions are considered in one integral model. It appears, however, at present to be difficult to collect the explicit human performance data that are needed for implementation of the model.

This kind of integrated model has great importance for design of aviation and other vehicle systems because the decision and manual control tasks of the operator form one integrated task in direct coupling with system dynamics. The task of a pilot or driver is a direct space-time control of a moving physical object, the vehicle. The sensorimotor level of the human information processing in this task serves for control signal processing; i.e., the output manual actions are continuous signals in the semiotic sense.

For process plants of the present levels of automation the continuous control signal processing is, however, automated. This means that human output actions will typically be related to switching and valving, and will be interpreted as stereotyped signs by the plant systems. This means that the human sensorimotor behavior will largely be used for an interface manipulation skill. For this, the time-space characteristics will have no direct relation to the basic system dynamics or the supervisory control tasks for which they in a way act as separating interface. For such systems, the interface manipulation skill is probably most conveniently described separately with reference to the parametric description in terms of the gain-bandwidth-accuracy mentioned above. In particular, the state identification can only be represented in terms of a Kalman observer when the dynamic properties of the system to be controlled are known in terms of the parameters of control theoretic state-space description. This is only the case for well-structured systems as, for instance, aircraft and space vehicles, and to some extent, the internal thermodynamic processes of industrial plants, and not for object manipulation in the physical environment in general. In consequence, models based on optimal control theory are only suited to represent the state identification of sensorimotor behavior, and the feature formation necessary to release and modify skilled patterns in case of manual control of well-structured dynamic systems, from which information is interpreted as signals. In less-structured situations when feature formation and state identification are based on recognition of information interpreted as signs, models of human judgment in terms of statistical representations are more suitable.

Models of Human Judgment

State identification and recognition based on a pattern of cues or signs are typical front-end functions involved in the behavior at the rule-based level. Patterns of information from the environment–or the internal representation of the problem space–are interpreted as signs referring to manual acts or information processes. The process is not based on functional or symbolic reasoning, and the association from sign to response may be based on purely empirical evidence from prior trials or learned from a teacher. The person will typically not be able to identify the information from the environment acting as cues for his decisions, and very probably conscious introspection may only lead to invalid after-rationalization, because the basis of introspection is a process of a higher cognitive domain than involved in the rule-based behavior itself.

Figure 12.2. Different approaches to models of human decision and judgment and their relations to economical and psychological concepts. [Adapted from Hammond et al. (1980) with permission from Hemisphere Publishing.]

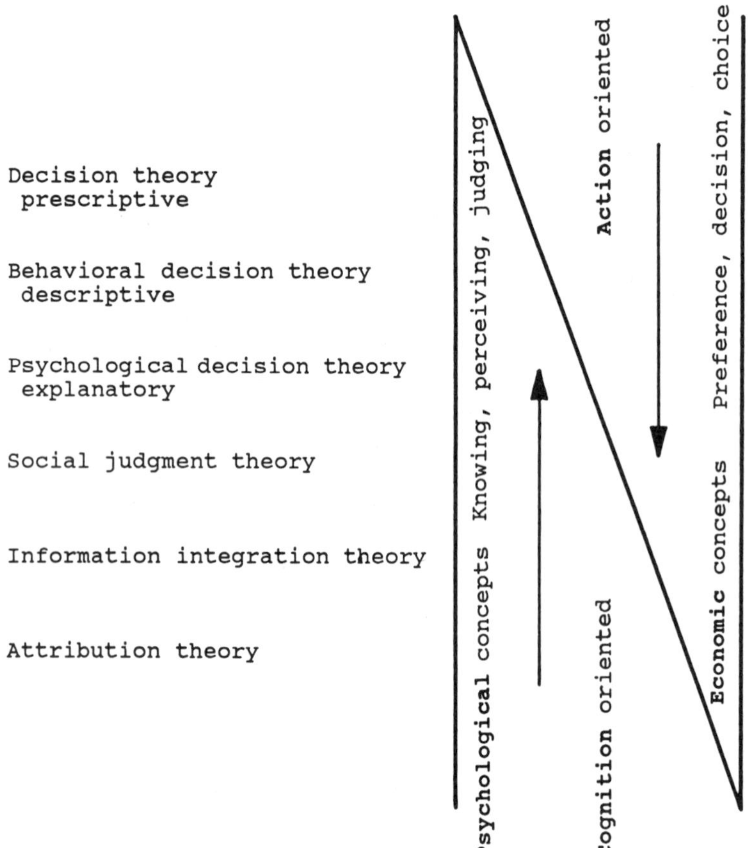

Models of judgment and decision making in everyday professional life at the rule-based level have very important implications for all trades, in particular for education and training and for the design of supporting tools. Therefore, models of this process have been the topic of extensive research in the area called human decision making, judgment, or choice, depending upon the paradigm of the approach. Several different approaches can be identified in this modeling effort, ranging from purely normative models based on economic concepts to predictive models derived from psychological research. Hammond et al. (1980) have recently reviewed these different approaches, and their classification has been adopted in the following section. The varying dependence of the theories on economic and psychological concepts is illustrated in Figure 12.2. In varying degrees, the theories consider the problem as a one-sided cognitive problem or a two-sided problem also involving the task system in the environment (see Figure 12.3). In the following section, the various models will be characterized briefly.

Decision Theory

This model is a mathematical model based on the expected utility theory developed by economists (Morgenstern) and mathematicians (von Neumann). The theory can be traced back to Bernoulli's work on the worth of a decision determined by the probability of events and their associated utilities. Decision theory limits its interest to the single-system case, which involves one person without full knowledge of the task situation and without feedback about the effect of the decision. The approach focuses on decision making from a prescriptive point of view only. It is a logical structure for decisions and makes no claim that it represents or describes the information processing of human decision makers. The emphasis is not on what they do, but what they should do. Modern theorists (Keeney and Raiffa, 1976) emphasize the mathematical modeling of subjective probability and utility and promote the use of the theory to aid decision makers to achieve logical consistency.

Critics of decision theory argue, however, that it is not useful as a guide because human beings do not behave in accordance with the fundamental assumptions of the theory. As it is a normative model, it will not be considered in more detail in the present context. It may, however, be an important candidate to consider for computer implementation in decision support systems.

Behavioral Decision Theory

This theory was initially developed by Edwards and Tversky (1967) and is based on the Bayesian probability theory and on an experimental approach to modeling. The aim of the experiments has been to find out how closely human information processing approximates the Bayesian process, and how information is used to revise subjective probabilities. Experimental manipulations of the objective probabilities were used to examine revisions of the subjective

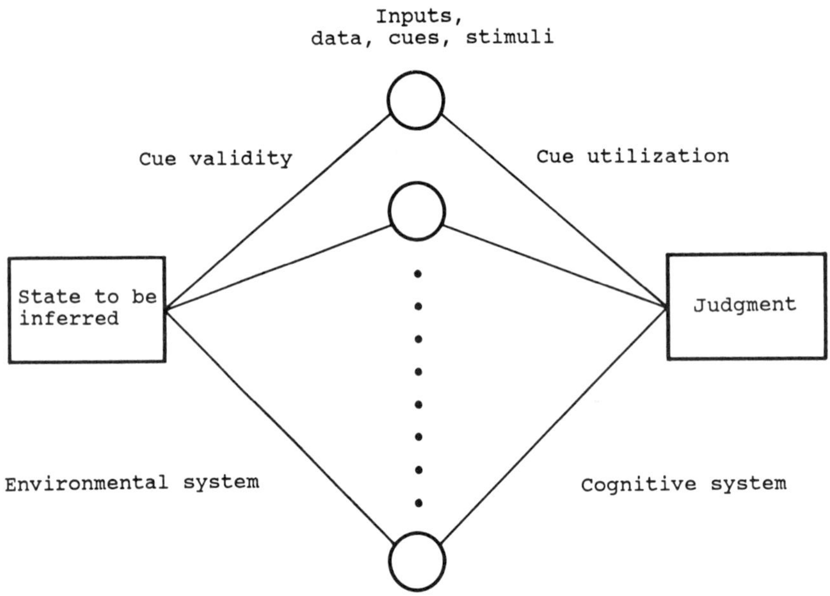

Figure 12.3. Decision theories consider the cognitive system and the environmental system to varying degrees. [Adapted from Hammond et al. (1980) with permission from Hemisphere Publishing.]

probabilities. Because decision tasks with known properties are included in the models, the theories consider the double-system case of Figure 12.3.

The approach intends to describe human departures from optimal performance empirically and to explain such departures in terms of both the external (task) and the internal (psychological) conditions. An often-mentioned departure from optimal decision making is conservatism, which represents the failure of humans to revise their posterior subjective probability as much as they should upon the receipt of new information. People are "conservative Bayesians."

Behavioral decision theory and classic decision theory share the same basic concepts as decision, preference, subjective probability, and utility, and both refer to the cognitive process of continuing probability and utility as "aggregation." These concepts are inherited from economic theory and mathematics, not psychology. However, unlike classic decision theory, behavioral decision theory does not aim at a prescription of the decision process, but at a description of the actual deviations from the optimal process. As such, it has had a rather extensive use to describe deviation from rational behavior in experiments on diagnostic behavior, for instance, in electronic troubleshooting (Rigney et al., 1968).

Psychological Decision Theory

In its recent form, this theory is based on the work of Tversky and Kahneman (1974, 1979) with their concepts of representativeness, availability, and the use of heuristics. Their approach rejects the use of optimal decisions as a frame of reference for description. It turns instead to a search for the psychological mechanisms that people use to evaluate frequencies and likelihoods. Psychological decision theorists generally consider subjective utility theory to be empirically unsuccessful and to be replaced by a theory that *explains* human behavior, that not only explains *why*, but is also able to predict *when* people replace the laws of statistics by heuristics. The theory is still rooted in decision theory, using probabilities and utilities as central descriptive forms, but focuses on the way in which people assign probabilities to events on explicit formulation of the *biases*, and upon classification of means for supporting decisions by *debiasing* (because it appears that simply warning people against their bias often proves ineffective).

Tversky and Kahneman have identified a number of biasing heuristics that are relevant also for the human-machine context (Tversky and Kahneman, 1974). Because the basic role of the heuristics is "to reduce complex tasks to simpler judgmental operations," they are relevant in many situations during disturbed plant operation.

The heuristics and their most important biasing effects are as follow:

Representativeness. People associate to a prototyped number of a class, neglecting prior probabilities and base-rate frequencies as well as the effect of sample sizes, and are influenced by the illusion of validity as well as by the misconception of regression toward the mean.

Availability. People assess the frequency of a class or the probability of an event by the ease with which instances of their occurrence could be brought to mind. They are typically influenced by bias owing to easy retrieval of familiar and recent events, and by bias owing to the effectiveness of a search set and the imaginability of situations, independent of probability.

Adjustment and anchoring. People make estimates from an initial value that may be suggested by the formulation of a problem and that is generally not adequately adjusted.

The heuristics and bias must be viewed as characteristics of the cognitive process, not as the effects of emotional or motivational behavioral factors (as, for example, wishful thinking). "The main cause for the failure to develop valid statistical intuitions is that events are normally not coded in terms of all the factors that are crucial to the learning of statistical rules" (Tversky and Kahneman, 1974); i.e., they operate from convenient signs, not defining attributes, as is discussed in Chapter 11.

Social Judgment Theory

This theoretical development as well as the basic concepts of "ecological validity of cues" and "utilization of cues" originate in Brunswik's theory of perception. The primary intention of the approach is not to explain, but to *describe* human judgment processes, and to provide guides for the development of decision aids. Central to the development has been the work of Hammond, Brehmer, and others (see Hammond et al., 1980). The basic framework of the theory is Brunswik's "lens model" (see Figure 12.4). An important feature of Brunswik's theory of perception is that he distinguishes clearly between the term for input of the S-R (stimulus-response) psychologist, which is "stimulus," pointing from the environment toward the person, and the term "cue," pointing outward from the person toward the aspects of the environment. Central to the model, as shown in Figure 12.4, is that the environment and the person are described in symmetric terms, hence the name "lens model." The cues of the task vary in "ecological validity" and the person has a variation in "cue utilization," both of which may assume various linear and nonlinear forms. Both sides are analyzed and attempts are made to match ecological validities to cue utilization, as well as ecological

Figure 12.4. Brunswik's "lens model." [Adapted from Hammond et al. (1980) with permission from Hemisphere Publishing.]

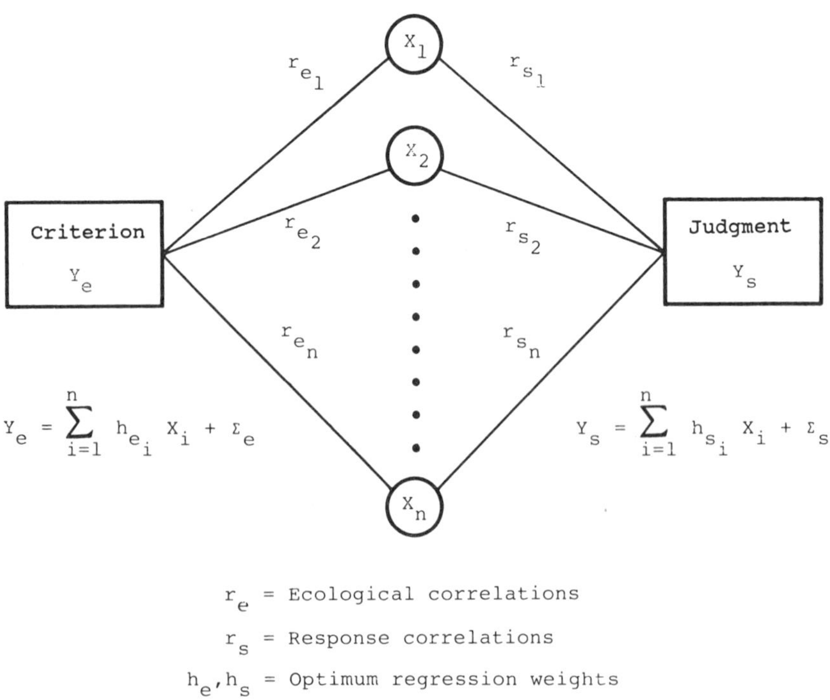

function forms to subjective function forms, i.e., to identify the extent to which the principles of organization that control the task system are reflected in the principles of the organization that control the cognitive system of the person.

The "lens model" has been the basis of research concerning diagnostic judgments in several professional activities such as, for instance, stockbrokers, clinical psychologists, and physicians (Brehmer, 1981). A problem in such research is to describe a mental process of which the person is not himself aware. The approach has been to assume that even though the person is not aware of the process, he will know the information, i.e., the cues, on which the process is based. In experiments, therefore, cues identified as diagnostically relevant by expert judges are used to present, generally in written form, subjects with trial cases. From the experimental research, the statistical model describing diagnostic behavior is identified. The general result has been that linear statistical models, such as multiple regression analysis, have been adequate. Four general results are typical of such diagnostic experiments. First, the judgment process tends to be very simple. Even though persons identify up to 10 cues to be relevant to diagnosis, they actually use very few, usually only two or three, and the process tends to be purely additive. Second, the process tends to be inconsistent. Subjects do not use the same rule from case to case, and judgment in a second presentation of a case may differ considerably from what it was the first time. Third, there are wide individual differences even among subjects with years of experience. They differ with respect to the cues used and the weights they apply. The fourth general result is that people are not very good at describing how they make judgments (Brehmer, 1981).

Given that intuitive judgments of an expert are performed by processes below the level of conscious attention, one may wonder whether laboratory experiments based on formal cues identified consciously by professional experts (textbook cues?) really correlate with the cues used in the real-life situation in which cues depending upon prehistory and informal "situational cues" may play an important role.

The review of Hammond et al. (1980) considers two further schools of theorists with strong relation to psychological concepts and methods: the "information integration theory" and the "attribution theory," which were developed within social psychology and have a strong emphasis upon interpersonal judgments. They have not, to my knowledge, been applied to human-machine systems but should probably be considered for group decision-making situations. In the following section, they will be characterized briefly.

Information Integration Theory

This theory is based on the work of Anderson (1974) and has its origin in psychophysics. In contrast to the previously mentioned theories, it is not a probabilistic theory. The aim of the theory is to discover psychological laws that intervene between stimulus and response in quantitative terms of a "cognitive algebra" (Anderson, 1974). It focuses on the organization and integration of information by means of various algebraic formulations such as additive equations, averaging equations, etc., which are considered models of cognitive functions. Discussing social judgments, Anderson (1974, p. 84) states: "Cognitive integration seems to follow simple averaging, subtracting, and multiplying rules far more commonly than has been recognized." Central to the theory is "functional measurement," which is strongly related to psychophysical methods of measuring complex social rather than physical data. Information integration theorists criticize social judgment models for only considering the subjects' physical stimuli, while information integration also considers conditions employed in traditional psychophysics such as, for example, social effects, anchoring, contrast, etc. There has been no interest within this theory to evaluate the fit of empirical behavior to models of optimal behavior; only models that account for the subjects' actual behavior are considered.

Attribution Theory

The basis of this theory is the theoretical work of Heider (1958) but has been developed by several researchers, for instance Kelly (1973), who has given several survey papers. The theory has been central within social psychology. The theory is very general in its aim and tends to cover not only a single person making judgment of the cause of an event in a social situation but also interpersonal relations in judgment, learning, and conflict. The central concept of the theory–"attribution"–can be considered a special case of inference or judgment. Choice of action or preference for friends, for instance, follow from different attributions, but the theory is primarily concerned with inferences about causality, i.e., causal attributions, and should therefore probably be considered more in regard to human-machine systems than it presently is.

Kelly (1973) considers two different cases. First, the attributor has information from several observations and is able to respond to covariation between the observed effect and its possible causes. Second, the attributor has information from only a single observation and must respond to the set of conditions present at a given time. Thus, it is necessary for him to take account of the configuration of factors that are plausible causes. By the covariation concept, an effect is attributed to one of its possible causes with which it covaries over time. Implicit in this principle is the problem of the exact temporal relations assumed to exist between a cause and its effect.

Kelly and Heider conceptualize the attribution process in terms of the analysis of variance as employed by the psychologist to interpret experimental results. "The assumption is that the man in the street, the naive psychologist, uses a naive version of the method used in science" (Kelly, 1973, p. 109). In this analysis of variance, the salient possible causes constitute the independent variables and the effect constitutes the dependent variable. For a wide range of

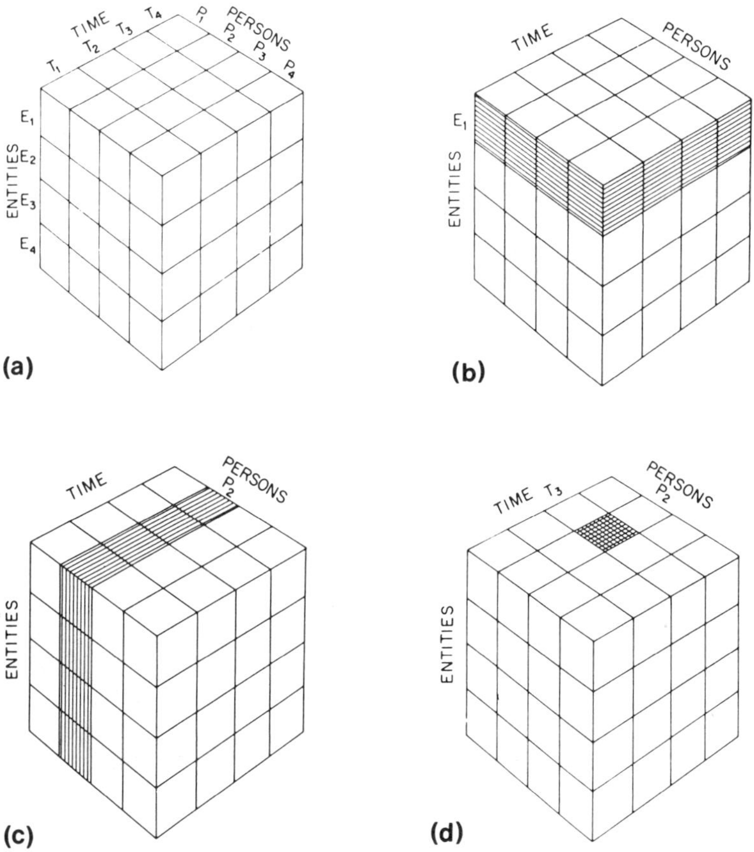

Figure 12.5. Analysis of variance of attribution theory. (a) The analyses of variance framework for making causal inferences; (b) data pattern indicating attribution to the entity; (c) data pattern indicating attribution to the person; (d) data pattern indicating attribution to the circumstances. [Reproduced from Kelley (1973) with permission from the American Psychological Association.]

attribution problems, the classes of possible causes are as shown in Figure 12.5: persons, entities, times. An important application of the persons *x* entity *x* time framework has to do with attribution validity. A response is known to be valid if (a) the response is distinctively associated with the stimulus, (b) there is consensus, other people have a similar response, and (c) the response is consistent over time. These criteria have been tested experimentally, and studies designed to test the idea that naive subjects treat information informally in a manner similar to the way statisticians treat it formally have been supportive. *Configuration concepts* are the basis of causal inference from a single observation of the effect. A person is rarely acting in complete ignorance, in that

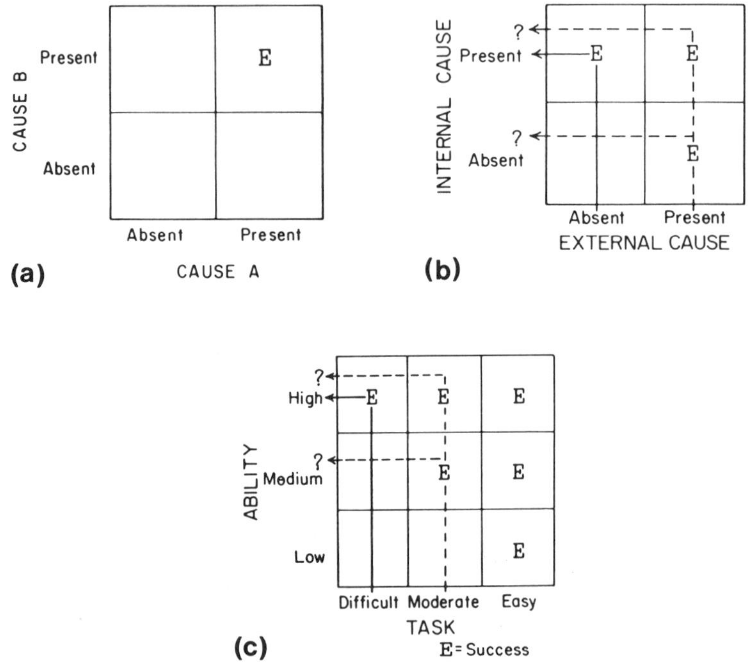

Figure 12.6. Causal schemata of attribution theory for (a) multiple necessary causes, (b) multiple sufficient causes, and (c) compensatory causes. [Reproduced from Kelley (1973) with permission from the American Psychological Association.]

ordinarily he has observed similar effects before and has some notions about possible relevant causes. Several statements can be made about the way attributors think: the discounting principle, stating that a given cause in producing a given effect is discounted if other plausible causes are also present; the augmentation principle, stating that if the external cause tends to inhibit or suppress the observed effect, the presence will increase the attribution of causes internal to the person observed, etc. The configuration concept has been implemented in a set of causal schemata related to compensatory causes, multiple necessary causes, and multiple sufficient causes (see Figure 12.6). Such causal schemata provide a person with means for making causal attributions given only limited information. People have repertoires of causal schemata enabling them to deal with causal problems, and prototyped features can be identified. A major task for attribution theory is to specify when a given schema is evoked. In this function also, stereotypes can be found: tendencies to prefer simple, e.g., single-cause patterns. This tendency has also been identified by Duncker (1945). He notes that certain cause-and-effect connections are intelligible to an attributor without evidence of covariation: a track resembles a foot of an animal; heavy things make loud noises, etc. Another simplification is that consequences will be linked directly to an actor rather than being viewed as a joint function of him and the situation.

As with other theories, systematic discrepancies between normative and intuitive inferences have been identified: subjects fail to extract all available information; they do not gain as much confidence from a series of events characterized by consistency as a probability model suggests they should, or they give too much weight to exceptions from general trends. A tendency to find causal explanations for variations that should be attributed to sampling variability is related to the bias from representativeness as discussed by Tversky and Kahneman (1974).

Attribution theory has, as mentioned, not been much considered in the context of human-machine systems, but the results obtained should probably be considered, for instance, in relation to group decision and personnel management modeling and in relation to attribution of the cause of an incident to personnel error. Finally, as control systems grow increasingly complex and functionally intransparent, the operator-control system relationship becomes more of an interpersonal relationship and, consequently, a social-psychological approach such as attribution theory may be an appropriate point of view.

Fuzzy Set Theory

In the typical models of human judgment, the relationship between the criterion and the attributes as well as between the attributes is considered to be stochastic in nature. This means that the categories and attributes are considered to be well defined and describable by classic logic including the rule of exclusion of the mean; i.e., an attribute is present or not present, and a judgment is either true or not true with a certain probability. The uncertainty may be due to the fact that the underlying phenomena are stochastic by nature, but it may also be due to lack of knowledge on the part of the observer, to choice of attributes that are convenient cues rather than defining attributes, or may be caused by discrete representation in verbal statements of states of affairs that are in reality overlapping and fuzzy. This had led to extensive efforts to develop models of fuzzy reasoning (Gaines, 1976) based on Zadeh's fuzzy set theory (Zadeh, 1965). An extensive bibliography on applications can be found in Gaines and Kohout (1977). The idea is basically that membership of a category is defined by a continuous membership function varying in the interval 0-1, for instance, the membership value of a temperature related to the category "hot" is a continuous function of measured temperature; the membership value of a man related to the class "tall" is a continuous function of his height around 180 cm. This means that the classic probability distributions are replaced by "possibility distributions." From the literature it seems that two different developments can be distinguished. One is the use of fuzzy sets in familiar many-valued logic; another is development of a fuzzy logic in which the truth values are themselves fuzzy sets. Doubts have been raised about the value of the latter approach (Haack, 1979) and, in general, the value of fuzzy sets compared with more traditional approaches seems to be a bit obscure.

Experimentally verified fuzzy models have typically been based on fuzzy sets that have been manipulated using nonfuzzy operations, which could have been handled by the traditional sum-of-weighted-attribute decision models. King and Mamdani (1977) have studied a fuzzy set model of a simple manual control task. Using verbal protocols, they identified the following human operator control algorithms: if the error is "positive medium" or "big," and if its rate of change is "negative small," then the control input should be "negative medium." The control action to choose from their observations was then selected from the possible actions as the one having the highest composite membership value. The individual membership grades were then quantified heuristically so that the performance of the algorithm was optimized. Comparison with the performance of an algorithm based on conventional control design for control of practical systems proved that the fuzzy approach yielded better performance. A model based on a fuzzy set representation of human cement kiln operators' heuristic control rules has been developed by Umbers and King (1980) and afterward implemented in an automatic control system. They found it possible to obtain a satisfactory control, but experienced difficulties in having operators explain their behavior to a degree that made it possible to reach an adequate fuzzy model. The basic advantage of the fuzzy set approach seems to be its combatibility with ambiguous verbal statements.

Fuzzy set theory has been used by Hunt and Rouse to model the diagnostic behavior in fault finding tasks (Rouse and Hunt, 1981; Hunt and Rouse, 1984). They used a two-level model based on the distinction between symptomatic and topographical search, discussed in Chapter 5. The idea is that an expert will operate on the rule-based level by associating symptoms to proper repair acts as long as he finds them useful. When this is no longer the case, he will turn to the knowledge-based topographical search based on the functional topology of the system. The definition of an expert then depends on the ability to realize when his expertise is no longer valid. This aspect will be discussed in more detail in relation to "Expert Systems" in the next section.

Hunt and Rouse (1984) define some symptomatic and topographic search rules that are derived from interviews of repair staff and laboratory experiments. The problem attacked by fuzzy set theory is which rule to apply in a given situation. In order to apply the knowledge contained in the rules, an algorithm for selecting rules is needed. Experience gained through prior diagnosis experiments has led Hunt and Rouse to the conclusion that four factors appear to significantly affect the rule selection process. For a rule to be selected: (a) the rule must be recalled; (b) the rule must be applicable to the current situation; (c) the rule must have some expected usefulness; and (d) the rule must be simple. It is hypothesized that rules are chosen on the basis of these four attributes. In most realistic situations, it is not always easy to make unambiguous assessments of the usefulness of a given rule. Further, from a human problem-solving point of view, recall and applicability are not simple either-or attributes. These attributes can therefore be considered to be fuzzy sets. In the model, the choice of the rule to apply in each instance is based on an evaluation of each of the available rules according to its membership value in the sets of recalled, applicable, useful, and simple rules. The model has been tested experimentally, and the approach seems to be promising (Hunt and Rouse, 1984). An interesting feature of this model may be its potential for learning the membership functions by experience (Rouse and Hunt, 1981).

Artificial Intelligence Models; Scripts and Plans; Expert Systems

During the recent decades there has been an extensive effort to design "artificial intelligence" systems based on analysis of human information-processing strategies. Different points of view concerning the aims of this development have been expressed, varying from design of intelligent machines that consider only human behavior as a source of design ideas, to the view that AI programs are to be taken as models of human performance (Ringle, 1979). Taken with caution, i.e., viewing AI programs as one possible implementation of a cognitive model without necessarily an isomorphic relation between elementary processes of human and computer, the cognitive models of AI are at present the most promising approach to simulation of higher-level processes in system design.

Development within artificial intelligence has different typical periods and approaches; see, for instance, the review by Dreyfus (1981). The early period was mainly based on the general, context-free problem-solving abilities of humans; a typical example is the general problem-solver program (GPS) by Newell et al. (1960). This approach is tightly related to human behavior in the knowledge-based domain and is based on a representation of the basic causal structure of a problem, and will be discussed in more detail below. After some years of optimism, however, the development turned more toward models based on an extensive representation of domain-specific, procedural knowledge and of commonsense reasoning (Schank and Abelson, 1977). This means that the focus has shifted toward models of human behavior at the rule-based level, and typical for this approach are the various "expert systems" (Feigenbaum, 1979; Hayes-Roth et al., 1983). The knowledge representation in such systems is based on the "rules of thumb" used by human experts. These rules are not principles of general reasoning, and long inference chains that develop from general rules used on structural representations are rarely used by "expert systems." The basic deductions behind the rules of the real expert are not repeated; the systems have stored the symptom-based responses of an expert within a limited domain.

Recently there has been some concern about the reliability of such expert systems (Barnett, 1982). The symptom-based rules are derived from experts' experience rather than being model-based. Therefore, they are usually very effective and produce correct behavior, but they also have the potential to produce inconsistent responses to unfamiliar information. "The rules are plausible and work a high percentage of the time; this is why the expert uses them. However, when they fail, the human expert will know enough to realize this fact and find out why. He retreats to a better grounded model (one based upon more general principles)" (Barnett, 1982). In other words, the typical expert system only models human behavior at the rule-based level and lacks the ability to retreat to knowledge-based problem solving–it is only able to interpret information as signs. Therefore, the systems fail abruptly when the environment changes and no longer conforms with the experience behind the rules.

Problem-Solving Models: Artificial Intelligence Approaches

Problem solving at the knowledge-based level is, even now, best represented by the results of the classical analysis of verbal protocols by Selz (1922), Duncker (1942), de Groot (1963), etc. It is also typical that the approach taken by Newell et al. for simulating knowledge-based problem solving by means of production systems (Newell et al., 1960; Newell and Simon, 1972) is based on this research. Based on a representation of the problem in a well-formed problem space, i.e., a model of the basic structure of the problem, behavior is developed from a set of production rules that are general, context-independent inference rules. A major weakness of this approach has been the difficulty of representing the ability of humans to consider only promising lines of reference for further development. This intuitive ability to see "where the problem is" is explicitly excluded from de Groot's analysis of the verbal protocols, and is not present in the production type models; see Dreyfus' comment on this issue (Dreyfus, 1972).

Generally, AI models have severe limitations because of the difficulty in representing the intuition or context present in human thinking stemming from the "subconscious world model"–or as Dreyfus expresses it–because of the lack of a physical body (Dreyfus, 1972). However, AI models are even then the best available tool for simulation of human information processing, and the development is important, not only for design of intelligent information-processing robots, but also for representing the human part in simulation of human-machine systems.

Figure 12.7. The various models of human behavior are suitable for different functions of an overall model. An important problem is to model the interaction among the various domains of behavior.

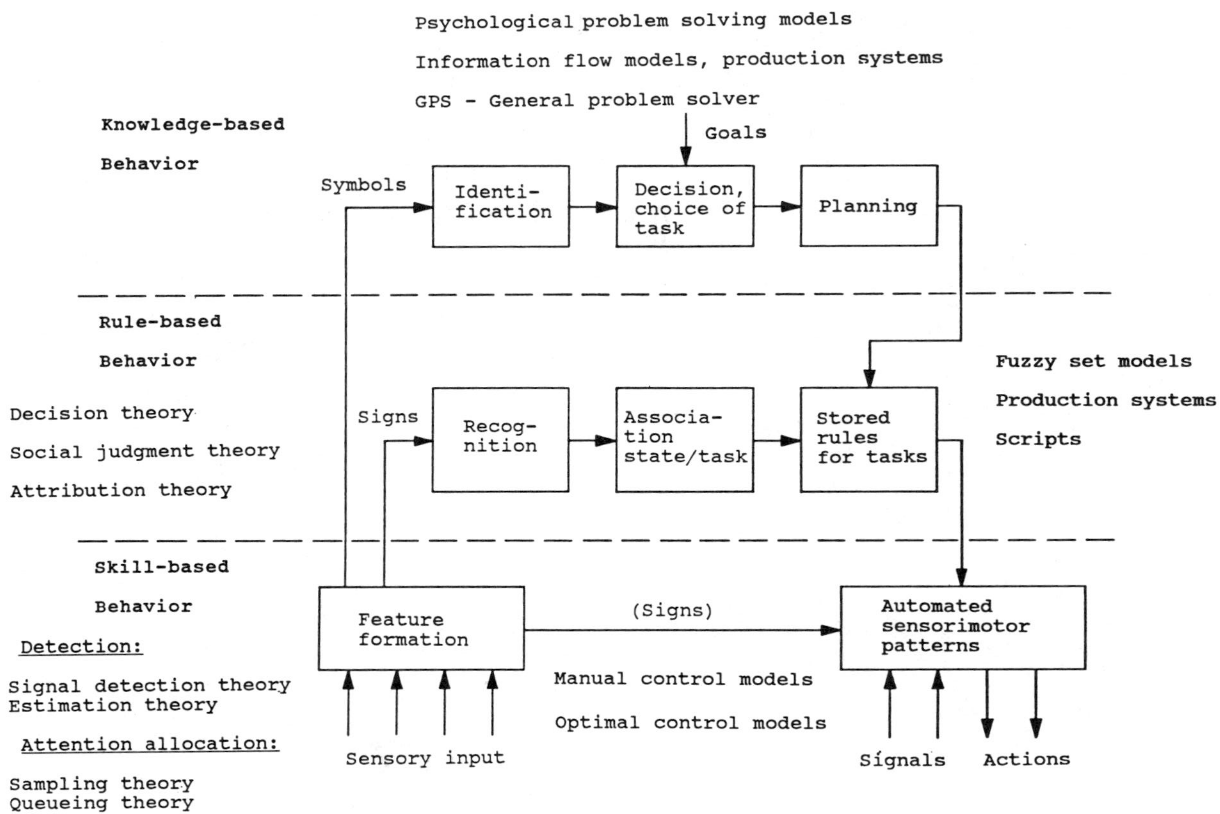
Psychological problem solving models
Information flow models, production systems
GPS - General problem solver
Knowledge-based
Behavior
Goals
Symbols
Identi-fication
Decision, choice of task
Planning
Rule-based
Behavior
Fuzzy set models
Production systems
Scripts
Decision theory
Social judgment theory
Attribution theory
Signs
Recog-nition
Associa-tion state/task
Stored rules for tasks
Skill-based
Behavior
Feature formation
(Signs)
Automated sensorimotor patterns
Detection:
Signal detection theory
Estimation theory
Attention allocation:
Sampling theory
Queueing theory
Sensory input
Manual control models
Optimal control models
Signals
Actions

Figure 12.7.

Chapter 13

Epilogue

In the Preface, it was mentioned that the aim of the book is to present a conceptual framework useful for bringing together the developments within several professional fields and making them applicable for human-machine systems design. The aim is not to propose solutions to specific problems. The discussions have therefore been kept at a rather general level in order to allow each individual reader to evaluate the implications within his domain of experience. At least two major lines of detailed analysis are necessary to bring substance into the framework for specific applications and thus to optimize the use of advanced information technology. One is the analysis of the control requirements of the particular systems and a systematic formal mapping of their functional means-end hierarchy, together with a cognitive task analysis of the decision tasks allocated operators and computers. Another line of analysis is the study of mental strategies and of human subjective preferences in the specific real-life tasks. This area raises the problem of experimental methods within psychology. We need reliable methods for probing cognitive control structures and mental models during real task performance and during experimental conditions. How can computers be used for the identification of patterns in the sequential data strings of verbal protocols and human-machine interaction records from experiments and real systems and thus make such data practically useful? Can association tests and categorization tasks be used for routine probing of the evolution of the mental models and strategies of experimental subjects during training and experiments? This apparently requires a new direction within experimental psychology to include complex experiments in the laboratory repertoire, and it requires that psychologists not only focus their interest upon the human but include detailed analysis of the human's task environment. This development was foreseen by Brunswik in 1952 when he advocated equal attention by psychologists to the real-life task content and to the psychological processes of the performer (see Brehmer, 1984).

This development within experimental psychology is crucial for the effective use of information technology in advanced human-machine systems, not only to develop the basis for systems design but, even more important, to develop methods and tools that allow a system designer–experimentally and analytically–to evaluate the match between his design intentions and the way the actual user(s) adapts to his system.

Bibliography

Ackermann, W., and G. Barbichon (1963). Conduites intellectuelles et activité technique. *Bull. CERP* 12 (1), 1-16

Alexander, C. (1964). *Notes on the Synthesis of Form.* Cambridge, MA: Harvard University Press.

Altman, J. (1979). Three levels of mentation and the hierarchic organization of the human brain, in *Psychology and Biology of Language and Thought: Essays in Honor of Eric Lenneberg*, G.A. Miller and E. Lenneberg, eds. New York: Academic Press, 87-109.

Anderson, N. (1974). Cognitive algebra: Integration theory applied to social attribution. *Exp. Soc. Psychol.* 7, 1-101.

Annett, J. (1981). Action, language and imagination. Conference on Memory, British Psychological Society, Sept 1981. Abstract in *Bull. Br. Psychol. Soc.*, 1982, May, Vol. 35.

Arnold, M.B. (1969). Human emotion and action, in *Human Action*, T. Mischel, ed. New York: Academic Press.

Atwoods, G. (1971). An experimental study of visual imagination and memory. *Cognit. Psychol.* 2, 290-299.

Baddeley, A.D. (1972). Selective attention and performance in dangerous environments. *Br. J. Psychol.* 63, 537-546.

Baron, S., and D.L. Kleinman (1969). The human as an optimal controller and information processor. *IEEE Trans. Man-Mach. Syst.* MSS-10, (1), 9-17.

Baron, S., G. Zacharias, R. Muralidharan, and R. Lancraft (1981). PROCHU: A model for analyzing flight crew procedures in approach to landing. *Proceedings of the Eighth World Congress of the International Federation of Automatic Control*, Vol. 7, 3481-3487. Oxford, England: Pergamon.

Baron, S., C. Feehrer, R. Muralidharan, R. Pew, and P. Horwitz (1982). An Approach to Modeling Supervisory Control of a Nuclear Power Plant. NUREG/CR-2988. ORNL/Sub/81-70523/1.

Barnett, J.A. (1982). Some issues of control in expert systems. *Proceedings of the International Conference on Cybernetics and Society*, Seattle, WA, October 28-30, 1982, 1-5.

Bartlett, F.C. (1943). Fatigue following highly skilled work. *Proc. R. Soc. Lond.* 131, 247-257.

Bartlett, F. (1958). *Thinking, An Experimental and Social Study*. London: Unwin.

Bateson, G. (1979). *Mind and Nature*. New York: Elsevier-Dutton Publishing Co.

Beishon, R.J. (1966). *A Study of Some Aspects of Mental Skill in the Performance of Laboratory and Industrial Tasks*. Ph.D. Thesis, University of Oxford. 293.

Bentley, A.F. (1950). Kennetic inquiry. *Science* 112, 775-783.

Berlyne, D. (1965). *Stucture and Direction in Thinking*. New York: John Wiley & Sons.

Bernstein, N. (1967). *The Coordination and Regulation of Movement* . New York: Pergamon Press.

Bosanquet, B. (1888). *Logic or the Morphology of Knowledge*. Oxford: Clarendon Press.

Bosanquet, B. (1920). *Implication and Linear Inference*. London: Macmillan.

Brehmer, B. (1981). Models of diagnostic judgments, in *Human Detection and Diagnosis of System Failures*, J. Rasmussen and W.B. Rouse, eds. New York: Plenum Press, 231-241.

Brehmer, B. (1984). Brunswikian psychology for the 1990s, in *Psychology in the 1990s*, K.M.J. Lagerspetz and P. Niemi, eds. New York: Elsevier Science Publishing Co.

Broadbent, D.E. (1971). *Decision and Stress*. London: Academic Press.

Brooks, L.R. (1967). The suppression of visualization by reading. *Q. J. Exp. Psychol.* 19, 289-299.

Brooks, L.R. (1968). Spatial and verbal components of the act of recall. *Can. J. Psychol.* 22, 349-368.

Brown, I.S., and J. DeKleer (1981). Towards a theory of qualitative reasoning, in *Human Detection and Diagnosis of System Failures*, J. Rasmussen and W.B. Rouse, eds. New York: Plenum Press, 317-335.

Bruner, J.S., J.J. Goodnow, and G.A. Austin (1956). *A Study of Thinking*. New York: John Wiley & Sons.

Bruner, J.S., R.R. Oliver, and P.M. Greenfield (1966). *Studies in Cognitive Growth*. New York: John Wiley & Sons.

Carbonell, J.R. (1966). A queueing model of many-instrument visual sampling. *IEEE Trans. Hum. Fact. Electron.* HFE-4 (4), 157-164.

Carbonell, J.R., J.L. Ward, and J.W. Senders (1968). A queueing model of visual sampling: Experimental validation. *IEEE Trans. Man-Mach. Syst.* MMS-9 (3), 82-87.

Cassirer, E. (1923). *Substance and Function.* Chicago: Open Court Publishing Co.; (1953). New York: Dover Publications, 1953.

Cassirer, E. (1944). *Essay on Man.* New Haven, CT: Yale University Press.

Craik, K.J.W. (1943). *The Nature of Explanation.* Cambridge: Macmillan.

Crossman, E.R.F.W., and J.E. Cooke (1962). Manual control of slow-response systems. International Congress on Human Factors in Electronics, Long Beach, California, 1962; Also in *The Human Operator in Process Control,* W. Edwards and F.P. Lees, eds. (1974). London: Taylor & Francis.

Cuny, X. (1971). *L'Approche Psycho-Semiologique: Étude d'un Code Gestuel de Travail.* Paris: Laboratoire de Psychologie du Travail de l'E.P.H.E. Report, November 1971.

Cuny, X. (1973). Les Commandements gestuels: Une éxperience avec ouvriers étrangers débutants. *Bull. Psychol.* 307, (26), 14-16.

Cuny, X., and M. Boy (1981). Analyse sémiologique et apprentissage des outils-signes: L'Apprentissage du schema d'electricit. *Communications* 33, 103-140.

Curry, R.E., and A.R. Ephrath. (1976). Monitoring and control of unreliable systems. in *Monitoring Behavior and Supervisory Control* , T.B. Sheridan and G. Johannsen, eds. New York: Plenum Press.

Cyert, R.M., and J.G. March (1963). *A Behavioral Theory of the Firm.* Englewood Cliffs, NJ: Prentice-Hall.

De Groot, A.D. (1965). *Thought and Choice in Chess.* The Hague: Mouton.

Dennett, D.C. (1971). Intentional systems. *J. Philos.* 68 (4), February 25, 1971.

Dreyfus, H.L. (1972). *What Computers Can't Do.* New York: Harper and Row; New revised version, 1979.

Dreyfus, H.L. (1981). From micro-worlds to knowledge representation, in *Mind Design,* John Haugeland, ed. Montgomery, Vermont: Bradford Books.

Dreyfus, S.E., and H.L. Dreyfus (1980). *A Five-Stage Model of the Mental Activities Involved in Directed Skill Acquisition.* Operations Research Center, ORC-80-2. University of California, Berkeley.

Duncker, K. (1945). On problem solving. *Psychological Monographs.* Vol. 58, No. 5, Whole No. 270.

Eastman, C.M. (1978). The representation of design problems and maintenance of their structure, in *Artificial Intelligence and Pattern Recognition in Computer-Aided Design,* J.-C.Latombe, ed. New York: North-Holland.

Eco, U. (1979). *A Theory of Semiotics.* Bloomington: Indiana University Press.

Edwards, W., and A. Tversky (1976). *Decision Making.* Baltimore; Penquin Books.

Ephrath, A.R., and L.R. Young (1981). Monitoring vs. man-in-loop detection or aircraft control failures, in *Human Detection and Diagnosis of System Failures* , J. Rasmussen and W.B. Rouse, eds. New York: Plenum Press.

Feigenbaum, E.A. (1979). Themes and case studies of knowledge engineering, in *Expert Systems in the Micro-Electronic Age*, D. Michie, ed. Edinburgh University Press.

Fitts, P.M., and M.I. Posner (1962). *Human Performance*. Monterey, CA: Brooks/Cole Publishing Co.

Forrester, J.W. (1971). *World Dynamics*. Cambridge, MA: Wright-Allen Press.

Furth, E., G. Grant, and H. Smithline (1967). *Data Conditioning and Display for Apollo Prelaunch Checkout*. NASA, Report N-68-12531. Darien, CT: Dunlapp and Associates.

Gai, E.G., and R.E. Curry (1976a). A model of the human observer in failure detection tasks. *IEEE Trans. Syst. Man Cybern.* SMC-6, (2), 85-94.

Gai, E.G., and R.E. Curry (1976b). Failure detection by pilots during automatic landings: Model and experiment. *J. Aircraft* 14 (2), 135-141.

Gaines, B.R. (1976). Foundations of fuzzy reasoning. *Int. J. Man-Mach. Stud* . No 8, 623-668.

Gaines, B.R., and L.J. Kohout (1977). The fuzzy decade: A bibliography of fuzzy systems and closely related topics. *Int. J. Man-Mach. Stud.* (9), 1-68.

Gasking, D. (1953). Mathematics and the world, in *Logic and Language*, A. Flew, ed. Oxford: Basil Blackwell.

Gibson, J.J. (1966). *The Senses Considered as Perceptual Systems*. Boston, MA: Houghton, Mifflin.

Goldstein, I, and S. Papert (1977). Artificial intelligence, language, and the study of knowledge. *Cognit. Sci.* (1), 84-123.

Goldstein, M. (1969). Display aspects of algebra. *Psychol.* Rep. 24, 937-938.

Goodstein, L.P., J. Hedegaard, K. Söe Höjberg and M. Lind (1983). *The GNP Testbed for Operator Support Evaluation*. Roskilde, Denmark: Risö National Laboratory, Report No. M-2460.

Goodstein, L.P. (1984). *Information Presentation for Decision Making* . Roskilde, Denmark: Risö National Laboratory, Report No. M-2461.

Gordon, W.J.J. (1963). *Synetics: The Development of Creative Capacity*. New York: Harper and Row.

Greene, T.M., ed. (1929). *Kant Selections*. New York: Charles Scribner's Sons.

Haack, S. (1979). Do we need fuzzy logic. *Int. J. Man-Mach. Stud.* (11), 437-445.

Hadamard, J.L. (1945). *The Psychology of Invention in the Mathematical Field*. Princeton: Princeton University Press.

Hamilton, P., G.R.J. Hockey, and M. Rejman (1977). The place of the concept of activation in human information theory, in *Attention and Performance*, Vol. VI, S. Dornic, ed. New York: Halsted Press.

Hammond, K.R., G.H. McClelland, and J. Mumpower (1980). *Human Judgment and Decision Making*. New York: Hemisphere Publishing, Frederick A. Praeger.

Hanford reactor incident (1955). Current events. *Nuclear Safety*. 1962, Vol.4, (1), 104-106.

Harman, G. (1976). Practical reasoning. *Rev. Metaphys.* 24 (3), 431-463.

Harman, G. (1982). Logic, reasoning, and logic form, in *Language, Mind, and Brain*, T.W. Simon and R.J. Scholes, eds. Hillsdale, NJ: Lawrence Erlbaum & Assoc.

Hayes-Roth, F., D.A. Waterman, and D.B. Lenat (1983). *Building Expert Systems* . Reading, MA: Addison-Wesley Publishing Company.

Heidbreder, E. (1949). The attainment of concepts. VIII. The conceptualization of verbally indicated instances. *J. Psychol.* 27, 263-309.

Heider, F. (1958). *The Psychology of Interpersonal Relations*. New York: John Wiley & Sons.

Held, R., and A. Hein (1963). Movement-produced stimulation in the development of visually guided behavior. *J. Comp. Physiol.* Psychol. 56, 872-876. Reprinted in (1970). *Perceptual Learning and Adaptation*, P.C. Dodwell, ed. Harmondsworth, England: Penguin Books, 502 pp.

Hockey, R. (1979). Stress and the cognitive components of skilled performance, in *Human Stress and Cognition*, V. Hamilton and D.H. Warburton, eds. New York: John Wiley & Sons.

Horowitz, M.J. (1967). Visual imagery and cognitive organization. *Am. J. Psychiatry* 123, 938-946.

Horowitz, M.J. (1970). *Image Formation and Cognition*. New York: Appleton-Century-Crofts.

Hubel, D.H. (1963). The visual cortex of the brain, *Sci. Am.* 209, 54-62.

Hunt, R.M., and W.B. Rouse (1984). A fuzzy rule-based model of human problem solving. *IEEE Trans. Syst. Man Cybern.* Vol. SMC-14 (1), 112-120.

Ihde, D. (1981). *Technics and Praxis. Dordrecht*, Netherlands: Reidel.

Janis, I.L., and L. Mann (1977). *Decision Making: A Psychological Analysis of Conflict, Choice, and Commitment*. London: The Free Press, Collier-Macmillan Publishers.

Jarvis, H.C. (1971). Butadiene explosion at Texas City—1. *Loss Prevention*. Vol. 5, 57-60.

Johannson, G. (1973). Visual perception of biological motion and a model for its analysis. *Percept. Psychophys.* 14, 201-211.

Johnson-Laird, P.N. (1980). Mental models in cognitive science. *Cognit. Sci.* 4, 71-115.

Kahneman, D., and A. Tversky (1979). Prospect theory: An analysis of decision under risk. *Econometrics* 47 (8), 263-291.

Kaufmann, G. (1979). *Visual Imagery and Its Relation to Problem Solving* . Oslo: Universitetsforlaget.

Keeney, R.L., and H. Raiffa (1976). *Decisions with Multiple Objectives, Preferences and Value Tradeoffs*. New York: John Wiley & Sons.

Kelley, H.H. (1973). The processes of causal attribution. *Am. Psychol.* Vol. 28 (2), 107-128.

King, P.J., and E.H. Mamdani (1977). The application of fuzzy control systems to industrial processes. *Automatica.* 13 (3), 235-242.

Kleinman, D.L., S. Baron, and W.H. Levison (1971). A control theoretic approach to manned-vehicle systems analysis. *IEEE Trans. Automat.* Control. AC-16, 824-832.

Kuhn, T. (1962). *The Structure of Scientific Revolutions*. Chicago: University of Chicago Press.

Lashley, K.S. (1942). The problem of cerebral organisation in vision, in *Visual Mechanisms*, H. Kluver, ed. Lancaster Cattell: (1969): *Biological Symposia*, 7, 301-322. Reprinted in *Perception and Action*, K.H. Pribram, ed. (Brain Behaviour, Vol. 2), 575. Harmondsworth, England: Penguin Books.

Less, F.P. (1981). Computer support for diagnostic tasks in the process industries, in *Human Detection and Diagnosis of System Failures*, J. Rasmussen and W.B. Rouse, eds. New York: Plenum Press, 369-388.

Leonard, T. (1960). Optimizing linear dynamics for human operated systems by minimizing the mean squared tracking error. *WESCON*, 4. (Pt. 4), 57-62.

Lettvin, J.V., H.R. Maturana, and W.S. McCulloch (1959). What the frog's eye tells the frog's brain. *Proc. IRE* 47, 1940-1951.

Lind, M. (1981). The use of flow models for automated plant diagnosis, in *Human Detection and Diagnosis of System Failures* , J. Rasmussen and W.B. Rouse, eds. New York: Plenum Press, 422-432.

Lind, M. (1982). *Multilevel Flow Modeling of Process Plant for Diagnosis and Control.* International Meeting on Thermal Nuclear Reactor Safety, Chicago. (Also: Roskilde, Denmark: Risö National Laboratory, Report No. M-2357).

Mach, E. (1905). *Knowledge and Error*. English edition, 1976. Netherlands: Dordrecht, Reidel.

Mackie, J.L. (1975). Causes and conditions, in *Causation and Conditionals*, E. Sosa, ed. Oxford: Oxford University Press.

Mandler, G. (1975). *Mind and Emotion.* New York: John Wiley & Sons.

Mandler, G. (1979). Thought processes, consciousness and stress, in *Human Stress and Cognition*, V. Hamilton and D.H. Warburton, eds. New York: John Wiley & Sons.

Michael, C.R. (1969). Retinal processing of visual images. *Sci. Am.* 220 (5), 105-114.

Michotte, A.E. (1963). *The Perception of Causality*. New York: Basic Books.

Miller, G.A. (1951). *Language and Communication.* New York: McGraw-Hill.

Miller, G.A. (1956). The magical number seven plus or minus two: Some limits on our capacity for processing information. *Psychol. Rev.* 63, 81-97.

Minsky, M. (1975). A framework for representing knowledge, in *The Psychology of Computer Vision,* P. Winston, ed. New York: McGraw-Hill.

Mintzberg, H. (1973). *The Nature of Managerial Work.* New York: Harper and Row.

Moray, N. (1981). The role of attention in the detection of errors and the diagnosis of failures in man-machine systems, in *Human Detection and Diagnosis of System Failures,* J. Rasmussen and W.B. Rouse, eds. New York: Plenum Press.

Morris, C. (1971). *General Theory of Signs.* Paris: Mouton.

Muralidharan, R., and S. Baron (1980). DEMON: A human operator model for decision making, monitoring and control. *J.Cybern. Inf. Sci.* (3), 97-122.

Neisser, U. (1972). Changing conceptions of imagery, in *The Function and Nature of Imagery,* P.W. Sheehan, ed. New York: Academic Press.

Newell, A., J.C. Shaw, and H.A. Simon (1960). Report on a general problem-solving program for a computer. *Information Processing: Proceedings of the International Conference on Information Processing.* Paris: UNESCO, 256-264.

Newell, A., and H.A. Simon (1972). *Human Problem Solving.* Englewood Cliffs, NJ: Prentice Hall.

Norman, D.A. (1980). *Errors in Human Performance.* Report No. 8004. University of California, San Diego: Center for Human Information Processing.

Norman, D.A. (1981). *Steps toward a Cognitive Engineering: Systems Images, System Friendliness, Mental Models.* Paper presented at Symposium on Models of Human Performance, Office of Naval Research Contractors' Meeting, La Jolla, CA, University of California, San Diego June 19, 1981.

Norman, D.A., and D.G. Bobrow (1975). On the role of active memory processes in perception and cognition, in *The Structure of Human Memory,* C. Cofer, ed. San Francisco: W. H. Freeman & Co.

Paivio, A. (1971a). *Imagery and Verbal Processes.* New York: Holt, Rinehart & Winston.

Paivio, A. (1971b). Image and language, in *Imagery, Current Cognitive Approaches,* S.J. Segal, ed. New York: Academic Press.

Pew, R.W. (1974). Human perceptual-motor performance, in *Human Information Processing: Tutorials in Performance and Cognition,* B. H. Kantowitz, ed. New York: Erlbaum.

Pew, R.W., D.C. Miller, and C.E. Feeher (1981). *Evaluation of Proposed Control Room Improvements Through Analysis of Critical Operator Decision.* Palo Alto, CA: Electric Power Research Institute, Research Project 891, NP-1982.

Polanyi, M. (1958). *Personal Knowledge.* London: Routledge & Kegan Paul.

Polanyi, M. (1967). *The Tacit Dimension.* New York: Doubleday & Co.

Pribram, K.H. (1966). Some dimensions of remembering: Steps towards a neuropsychological model of memory, in *Macromolecules and Behavior* . J. Gaito, ed. New York: Appleton-Century-Crofts, pp. 165-187.

Pribram, K.H. (1969). The Neurophysiology of Remembering. *Sci. Am.* 220 No. 1, 73-82.

Pribram, K.H. (1971). *Languages of the Brain.* Englewood Cliffs, NJ: Prentice-Hall.

Pylyshyn, Z.W. (1973). What the mind's eye tells the mind's brain. *Psychol. Bull.* 80 (1), 1-23.

Rabbit, P. (1979). Current paradigms and models in human information processing, in *Human Stress and Cognition*, V. Hamilton and D. M. Warburton, eds. New York: John Wiley & Sons.

Rasmussen, J. (1976). Outlines of a hybrid model of the process operator, in *Monitoring Behavior and Supervisory Control*, T.B. Sheridan and G. Johannsen, eds. New York: Plenum Press.

Rasmussen, J. (1979a). Notes on human error analysis and prediction, in *Synthesis and Analysis Methods for Safety and Reliability Studies*, G. Apostolakis and G. Volta, eds. London: Plenum Press.

Rasmussen, J. (1979b). *On the Structure of Knowledge–A Morphology of Mental Models in a Man-Machine Context.* Roskilde, Denmark: Risö National Laboratory, Report No. M-2192.

Rasmussen, J. (1980a). What can be learned from human error reports, in *Changes in Working Life*, K. Duncan, M. Gruneberg, and D. Wallis, eds. London: John Wiley & Sons.

Rasmussen, J. (1980b). The human as a system component, in *Human Interaction with Computers*, H.T. Smith and T.R.G. Green, eds. London: Academic Press.

Rasmussen, J. (1981). Models of mental strategies in process plant diagnosis, in *Human Detection and Diagnosis of System Failures*, J. Rasmussen and W.B. Rouse, eds. New York: Plenum Press, 241-258.

Rasmussen, J. (1982a). Human errors: A taxonomy for describing human malfunction in industrial installations. *J. Occupat. Accid.* 4 (2-4), 311-335.

Rasmussen, J. (1982b). Human factors in risk analysis, in *High Risk Safety Technology*, A.E. Green, ed. New York: John Wiley & Sons.

Rasmussen, J. (1983). Skills, rules, knowledge: signals, signs, and symbols and other distinctions in human performance models. *IEEE Trans. Syst. Man Cybern.* SMC-13 (3), 257-267.

Rasmussen, J. (1984). Strategies for state identification and diagnosis, in *Advances in Man-Machine Systems Research.* W.B. Rouse, ed. Greenwich, CT: J.A.I. Press.

Rasmussen, J., and A. Jensen (1973). *A Study of Mental Procedures in Electronic Troubleshooting.* Roskilde, Denmark: Risö National Laboratory. Report No. M-1582.

Rasmussen, J., and A. Jensen (1974). Mental procedures in real-life tasks: A case study of electronic troubleshooting. *Ergonomics* 17 (3), 293-307.

Rasmussen, J., and M. Lind (1981). *Coping with Complexity.* European Annual Conference on Human Decision Making and Manual Control, Delft. Also in Roskilde, Denmark: Risö National Laboratory, Report No. M-2293.

Rasmussen, J., O.M. Pedersen, A. Mancini, A. Carnino, M. Griffon, and P. Gagnolet (1981). *Classification System for Reporting Events Involving Human Malfunction.* Roskilde, Denmark: Risö National Laboratory, Report No. M-2240.

Ribeiro, S.T. (1967). Random-pulse machines. *IEEE Trans. Electron. Comput*. EC-16 (8), 261-276.

Reiger, C. (1976). An organization of knowledge for problem solving and language comprehension. *Artif. Intell.* 7, 89-127.

Reiger, C., and M. Grinberg (1976). *The Causal Representation and Simulation of Physical Mechanisms.* AD AO 34194, TR-49. College Park, MD: University of Maryland.

Rigney, J.W., D.M. Towne, and A.K. Mason (1968). *An Analysis of Structure and Errors in Corrective Maintenance Work.* TR-55, Electronics Personnel Research Group. Los Angeles: University Southern California.

Ringle, M. (1979). Philosophy and artificial intelligence, in *Philosophical Perspectives in Artificial Intelligence*, M. Ringle, ed. Sussex, England: The Harvester Press.

Roig, R.M. (1962). A comparison between human operator an optimum linear controller RMS-Error Performance. *IRE Trans. Hum. Fact. Electron.* HFE-3 (1), 18-21.

Rosch, E. (1976). Classification d'objects du monde réel: Origines et représentations dans la cognition. *Bull. Psychol.* Special Annual, 242-250.

Rosch, E. (1977). Human categorization, in *Advances in Cross-Cultural Psychology*, Vol. 1, N. Warren, ed. London: Academic Press.

Rosenbluth, A., N. Wiener, and J. Bigelow (1943). Behaviour, purpose and teleology. *Philos. Sci.* 10, 18-24.

Rouse, W.B. (1977). Human-computer interaction in multi-task situations. *IEEE Trans. Syst. Man Cybern.* SMC-7, (5), 384-392.

Rouse, W.B. (1980). *System Engineering Models of Human-Machine Interaction.* New York: Elsevier North-Holland, Inc.

Rouse, W.B. (1982a). A mixed-fidelity approach to technical training. *J. Educat. Technol. Syst.* 11 (2), 103-115.

Rouse, W.B. (1982b). Communication in panel on human performance theories and models. IEEE standards workshop on human factors in nuclear safety, in *Conference Record: The Man-Machine Interface and Human Reliability: An Assessment and Projection.* New York: IEEE.

Rouse, W.B. (1983). Models of human problem solving: Detection, diagnosis, and compensation for system failures, *Automatica* 19, 613-625.

Rouse, W.B., and R.M. Hunt (1981). A fuzzy rule-based model of human problem solving in fault diagnosis tasks. *Proceedings of the Eighth Triennial World Congress of the International Federation of Automatic Control.* Kyoto, Japan, August 1981.

Rouse, W.B., S.H. Rouse, and S.J. Pellegrino (1980). A rule-based model of human problem solving performance in fault diagnosis tasks. *IEEE Trans. Syst. Man Cybern.* SMC-10 (7), 366-376.

Rugg, H. (1963). *Imagination.* New York: Harper & Row.

Rubin, E. (1920). Vorteile der Zweckbetrachtung für die Erkentnis. *Z. Psychol.* 85, 210-223. Also in *Experimenta Psychologica,* 66-81. Copenhagen: Munksgaard.

Rumelhart, D.E., and D.A. Norman (1981). Analogical processes in learning, in *Cognitive Skills and Their Acquisition,* J.R. Anderson, ed. Hillsdale, NJ: Lawrence Erlbaum Associates.

Russel, B. (1913). On the notion of cause. *Proc. Aristot.* Soc. 13, 1-25.

Schank, R.S. (1975). *Conceptual Information Processing.* Amsterdam: North-Holland Publishing Co.

Schank, R.C., and R.P. Abelson (1977). *Scripts, Plans, Goals, and Understanding* . Hillsdale, NJ: Lawrence Erlbaum & Assoc.

Searle, J.R. (1981). Minds, brains, and programs, in *Mind Design,* J. Haugeland, ed. Montgomery, Vermont: Bradford Books.

Searle, J.R. (1983). *Intentionality.* Cambridge: Cambridge University Press.

Segal, S. (1971). *Imagery: Current Cognitive Approaches.* New York: Academic Press.

Segal, S. (1972). Assimilation of a stimulus in the construction of an image: The Perky effect revisited, in *The Function and Nature of Imagery,* P.W. Sheehan, ed. New York: Academic Press.

Selz, O. (1922). Zur Psychologie des produktiven Denkens und des Irrtums. Bonn: Friederich Cohen.

Senders, J.W. (1964). The Human Operator as a Monitor and Controller of Multi-Degree of Freedom Systems. *ERE Trans. Hum. Fact. Electron.* HFE-5, 2-5.

Senders, J.W. (1966). A reanalysis of pilot eye-movement data, *IEEE Trans. Hum. Fact. Electron.* HFE-7, 103-106.

Senders, J.W., and M.J.M. Posner (1976). A queueing model of monitoring and supervisory behavior, in *Monitoring Behavior and Supervisory Control,* T.B. Sheridan and G. Johannsen, eds. New York: Plenum Press.

Shacklee, H., and B. Fischhoff (1982). Strategies of information search in causal analysis. *Memory Cognit.* 10 (6), 520-530.

Shepherd, A., E.C. Marshall, and K.D. Duncan (1977). Diagnosis of plant failures from a control panel: A Comparison of three training methods. *Ergonomics* 20, 347-361.

Sheridan, T.B. (1981). Understanding human error and aiding human diagnostic behavior in nuclear power plants, in *Human Detection and Diagnosis of System Failures*, J. Rasmussen and W.B. Rouse, eds. New York: Plenum Press, pp. 19-35.

Sheridan, T.B. (1982). *Supervisory Control: Problems, Theory and Experiment for Application to Human-Computer Interaction in Undersea Remote Systems.* Department of Mechical Engineering, M.I.T. Technical Report, March 1982.

Sheridan, T.B., and W.R. Ferrell (1974). *Man/Machine Systems: Information, Control, and Decision Models of Human Performance.* Cambridge, MA: M.I.T. Press.

Sheridan, T.B., U. Tsach, and J. Tzelgov (1982). A new method for failure detection and location in complex dynamic systems. *Proceedings of the American Control Conference.* Arlington, VA, June 14-16, 1982.

Simon, H.A. (1969). *The Sciences of the Artificial.* Cambridge, MA: M.I.T. Press.

Sloman, A. (1971). Interaction between philosophy and artificial intelligence: The role of intuition and non-logical reasoning in intelligence, in *Proceedings of the International Artificial intelligence Conference.* London: The British Computer Society, pp. 270-278.

Sloman, A. (1978). *The Computer Revolution in Philosophy: Philosophy, Science, and Models of Mind.* Sussex, England: The Harvester Press.

Smith, H.T. (1976). Perceptual organization and the design of the man-computer interface in process control, in *Monitoring Behavior and Supervisory Control*, T.B. Sheridan and G. Johannsen, eds. New York: Plenum Press.

Sokolov, A.N. (1971). Internal speech and thought. *Int. J. Psychol.* 6, 79-92.

Sternberg, S. (1969). The discovery of processing stages. *Acta Psychol.* 30, 276-315.

Taylor. D.H. (1981). The hermeneutics of accidents and safety. *Ergonomics* 24 (6), 487-492.

Toynbee, A. (1972). *A Study of History.* London: Oxford University Press and Thames and Hudson.

Truxal, J.G. (1955). *Automatic Feedback Control System Synthesis.* New York: McGraw-Hill.

Tversky, A., and D. Kahneman (1974). Judgment under uncertainty: Heuristics and biases. *Science* 185, 1123-1124.

Umbers, I.G., and P.J. King (1980). An analysis of human decision making in cement kiln control and the implications for automation. *Int. J. Man-Mach. Stud.* 12, 11-23.

U.S. Congress, Oversight Hearings, Washington: (1979). *Accident at the Three Mile Island Nuclear Power Plant.* May 9-15, 1979. U.S. Gov. Printing Office.

von Foerster, H. (1966). From stimulus to symbol: The ecology of biological computation, in *Signs, Image and Symbol*, G. Kepes, ed. London: Studio Vista.

von Uexküll J., and G. Kriszat (1934). *Streifzüge durch die Umwelten von Tieren und Menschen, Bedeutungslehre.* Berlin: Springer. New edition (1956), Hamburg: Rowohlt.

Vygotsky, L.S. (1934). *Thought and Speech.* English Edition. Cambridge, MA: M.I.T. Press.

Wald, A. (1947). *Sequential Analysis.* New York: John Wiley & Sons.

Wason, P.C., and P.N. Johnson-Laird (1972). *Psychology of Reasoning.* Cambridge, MA: Harvard University Press.

Wertheimer, M. (1945). *Productive Thinking.* New York: Harper and Brothers.

Whitehead, A.N. (1927). *Symbolism, Its Meaning and Effect.* New York: Macmillan.

Whorf, B.L. (1940). Science and linguistics, in *The Technology Review,* Vol. XLII. Also in *Language, Thought and Reality. Selected Writing of Benjamin Lee Whorf,* J.B. Carroll, ed. (1956). Cambridge, MA: M.I.T. Press.

Wickens, C.D., and C. Kessel (1981). Failure detection in dynamic systems, in *Human Detection and Diagnosis of System Failures,* J. Rasmussen and W.B. Rouse, eds. New York: Plenum Press.

Williams, M.D., T.P. Moran, and J.S. Brown (1982). The role of conceptual models in nuclear power plant operations. *Proceedings of the Workshop on Cognitive Modeling of Nuclear Plant Control Room Operators.* Nuclear Regulatory Commission, Dedham, MA, August 1982.

Winch, P. (1958). *The Idea of a Social Science.* London: Routledge & Kegan Paul.

Wohl, J. (1981). System complexity, diagnostic behavior and repair time, in *Human Detection and Diagnosis of System Failures,* J. Rasmussen and W.B. Rouse, eds. New York: Plenum Press, 217-231.

Wohl, J. (1982). Information Automation and the Apollo Program: A Retrospective. *IEEE Trans. Syst. Man Cybern.,* Vol. SMC-12 (4), 469-478.

Young, L.R. (1969). On adaptive manual control. *Ergonomics* 12 (4), 635-675.

Zadeh, L.A. (1965). Fuzzy sets. *Inf. Control.* 8, 338-353.

Subject Index

E

F

Author Index